WHOLE
EARTH
DISCIPLINE

WHOLE EARTH DISCIPLINE

AN ECOPRAGMATIST MANIFESTO

───────────

STEWART BRAND

Atlantic Books
London

First published in 2009 in the United States of America by
The Viking Press, part of Penguin Group USA.

First published in Great Britain in hardback and airport and
export trade paperback in 2010 by Atlantic Books, an imprint
of Grove Atlantic Ltd.

1 3 5 7 9 10 8 6 4 2

A CIP catalogue record for this book is available from
the British Library.

Hardback ISBN: 978 1 84354 815 7
Trade Paperback ISBN: 978 1 84887 039 0

Printed in Great Britain by the MPG Books Group

Atlantic Books
An imprint of Grove Atlantic Ltd
Ormond House
26–27 Boswell Street
London
WC1N 3JZ

www.atlantic-books.co.uk

FOR JOHN BROCKMAN

------

who gives scientists
direct voice in the public discourse

CONTENTS

WHOLE
EARTH
DISCIPLINE

·1·
Scale, Scope, Stakes, Speed

We are as gods and HAVE to get good at it.

—*Whole Earth Discipline*

Climate change. Urbanization. Biotechnology. Those three narratives, still taking shape, are developing a long arc likely to dominate this century. How we frame them now will affect how they play out. Illusions abound on all three subjects, but their true nature is knowable.

In the face of climate change, everybody is an environmentalist. That's tough not just for people who have been comfortable thinking of themselves as antienvironmentalist; it's even tougher for long-time Greens. Activist Bill McKibben recently noted: "The environmental movement has morphed steadily into the climate change movement." That means that Greens are no longer strictly the defenders of natural systems against the incursions of civilization; now they're the defenders of civilization as well. It's a whiplash moment for everyone.

When roles shift, ideologies have to shift, and ideologies hate to shift. The workaround is pragmatism—"a practical way of thinking concerned with results rather than with theories and principles." The shift is deeper than moving from one ideology to another; the shift is to discard ideology entirely.

We are still realizing how much radical rethinking we will need to comprehend the forces now loose in the world and to figure out how to deal with them. The scale of forces, this time, is planetary; the scope is centuries; the stakes are what we call civilization; and it is all taking place at the headlong speed of self-accelerating human technologies and climatic turbulence. Talk of "saving the planet" is overstated, however. Earth will be fine, no matter what; so will life. It is humans who are in trouble. But since we got ourselves into this fix, we should be able to get ourselves out of it.

● The best way to think about climate change I found in a book that seems to be about something else—*Constant Battles* (2003), by Harvard archaeologist Steven LeBlanc (with Katherine Register). Drawing on abundant archaeological and ethnological evidence, LeBlanc argues that humans have always waged ferocious war. In all societies from hunter-gatherers on up through agricultural tribes, then chiefdoms, to early complex civilizations, 25 percent of adult males routinely died from warfare. No one wanted to fight, but they were constantly forced to choose between starvation and robbing the neighbors. Their preferred solution was the total annihilation of the neighbors.

The book is full of harsh revelations. Close examination of human burials shows that wholesale slaughter was common, and so was cannibalism—for nutrition, not ceremony. The abundant "cooking stones" at many archeological sites turn out to be ammunition—sling missiles (David killed Goliath with one). Dogs were the first animals to be domesticated because they make such good sentries, and that's why all dogs bark (and wolves don't). Most cities were walled.

Humans perpetually fight, LeBlanc says, because they always outstrip the carrying capacity of their natural environment and then have to fight over resources. Native peoples developed arcane knowledge of how to find and prepare difficult foods because they'd eliminated all the easy food sources. Peace *can* break out, though, when carrying capacity is pushed up suddenly, as with the invention of agriculture, or newly effective bureaucracy, or remote trade, or technological breakthroughs. Also a large-scale dieback from pestilence can make for peaceful times—Europe after its major plagues, the Americas after European diseases nearly obliterated the native populations. Such interludes are short: Population quickly rises to once more push against carrying capacity, and normal warfare resumes.

Only in the last three centuries, LeBlanc points out, have advanced states steadily lowered the overall body count to where just 3 percent of the world's people die from warfare these days, even though a few of the remaining wars and genocides have grown to world-war scale. Instead of butchering all their enemies in the traditional way, states merely kill enough to achieve a victory; then they put the survivors back to work. States also use their bureaucracies, advanced technology, and international rules of behavior to raise carrying capacity and sometimes even develop a careful relationship to it.

But all of that civilized sophistication could collapse if carrying capacities everywhere are lowered by severe climate change. Humanity would revert to its norm of constant battles for diminishing resources. Peace lovers would be killed and eaten by war lovers.

That's the prospect, I realized, reading LeBlanc. With climate change under way, we have to make a choice. If we do nothing or not enough, we face a carrying-capacity crisis leading to war of all against all, this time with massively lethal weapons and a dieback measured in billions. Alternatively, LeBlanc concludes,

> For the first time in history, technology and science enable us to understand Earth's ecology and our impact on it, to control population growth, and to increase the carrying capacity in ways never before imagined. The opportunity for humans to live in long-term balance with nature is within our grasp if we do it right. It is a chance to break a million-year-old cycle of conflict and crisis.

Up until 2003, I had only the usual concerns about climate change. Back in 1982, my wife and I bought an old tugboat to live on because it was impervious to the California hazards of earthquake and wildfire, and what the hell, because it was a cheap way to own a bayfront home with never a care about rising sea levels from global warming. Climate change was fun to think about, dire but distant.

I am employed half-time by a consulting company I helped found in 1987 called Global Business Network/Monitor (GBN). What happened in 2003 was that GBN got a request from the office of the U.S. Secretary of Defense to build a scenario about "abrupt" climate change. My role was peripheral; I did a few of the phone interviews with climatologists and contributed one idea. We delivered the report that fall—"An Abrupt Climate Change Scenario and Its Implications for United States National Security," by Peter Schwartz and Doug Randall.

Our scenario was based on an event that took place 8,200 years ago, when temperatures suddenly dropped 2.7° Celsius (5° Fahrenheit) in less than a decade. On the temperature charts, it's a one-century blip, nothing like the Younger Dryas event that humans endured 12,700 years ago, when world temperatures abruptly plummeted 15°C (27°F) and stayed that way for a thousand years. One explanation for both events is that the Gulf Stream was slowed or stopped by an excess of fresh water in the North Atlantic caused by global warming. (Data collected since 2003 variously challenge and support that theory; nobody denies that the violent climate events occurred).

Because our current global warming is melting Arctic ice and freshening North Atlantic waters, GBN's scenario examined what would happen if in, say, 2010 we entered an abrupt "minor" cooling like the one 8,200 years ago. The suddenly cooler, drier, windier world would experience droughts in the major agricultural areas, along with harsh winters and vicious storms and floods in unexpected places. By 2020, we said, Europe's climate would be more like Siberia's. Global food, water, and energy supply would be stressed. Earth's carrying capacity for humans would suddenly drop below what the 7.5 billion population of 2020 would require to survive. GBN concluded the report with my realization from LeBlanc, that whenever societies bash against carrying-capacity limits, they revert to the million-year-old norm of fighting over resources. By the 2020s, war, disease, and famine would be reducing human population until it came back into balance with the new carrying capacity. The Pentagon was an appropriate client for the scenario.

Deliberately kept unclassified, the report went public online and was summarized in *Fortune* magazine. At first a few keepers of the climate literature disparaged the scenario, but soon enough it became a widely cited part of that literature. The idea of abruptness (in our paper and a number of others) changed the public conversation about climate change. For the first time, climate was understood as a clear and present danger, the responsibility of currently serving officials worldwide instead of some future generation's problem. Public opinion on the subject began its own abrupt change.

● If GBN's scenario worries you, don't worry. In 2007 the Intergovernmental Panel on Climate Change (IPCC) consulted twenty-three climate models and concluded that the widespread concern of climatologists about the Gulf Stream was misplaced. A Norwegian professor, Helge Drange, said, "The bottom line is that the atmosphere is warming up so much that a slowdown of the North Atlantic Current will never be able to cool Europe." Or worry harder. A 2008 study of Greenland ice cores revealed that changes in the Gulf Stream appear to have triggered severe climate changes twelve thousand years ago that occurred not over decades but in one to three *years*.

Following climate science these days is a lot like the kid joke: Two men were flying in an airplane. Unfortunately one fell out. *Fortunately* there was a haystack. *Unfortunately* there was a pitchfork in the haystack. *Fortunately* he missed the pitchfork. *Unfortunately* he missed the haystack.

Fortunately, the IPCC climate models enabled thousands of scientists to publicly declare that global warming is real, that it is largely driven by human-generated greenhouse gases (CO_2 and methane mostly), and that the global consequences will become quite serious by 2040 and grow worse thereafter. *Unfortunately*, the IPCC models failed to predict the extremely rapid melting of Arctic ice—the ice was half gone in the summer of 2007 instead of the predicted 2050s.

In 2006 and again in 2008, Global Business Network ran a scenario workshop for the Arctic Marine Council on the future of shipping in the Arctic. I learned that sixty-five surface ships have been to the North Pole, that hundreds of cruise ships now visit the Arctic, that fish and fisheries are moving north, that the once-mythic Northwest Passage above Canada is opening for navigation, and that the Russians are pouring concrete for a series of ports along the Northern Sea Route, which will offer a shipping shortcut above all of Europe and Asia. The salty group in attendance at the first workshop—twenty-four assorted skippers, Coast Guard officers, polar scientists, diplomats, and CEOs—came up with four scenarios exploring the effects of various traffic demands and potential national and international policies concerning freedom of navigation, safety, and environmental protection in the new Arctic ocean passages. All four scenarios took as a given that the ice will keep right on melting.

● The reason is positive feedback. White ice reflects sunlight, 85 percent of it. Dark ocean absorbs sunlight, reflecting only 5 percent. Less ice leads to more absorption of solar heat in the Arctic, which leads to still less ice, which leads to ever more heat absorption, melting ever more ice: That's positive feedback. This particular phenomenon is called the ice-to-water albedo flip, *albedo* meaning reflectance.

Note that the word *positive* in the cybernetic term *positive feedback* does not mean "good." It usually means trouble, because it amplifies change. In the Wikipedia definition:

> *Positive feedback*, sometimes referred to as *cumulative causation*, is . . . a feedback loop system in which the system responds to perturbation in the same direction as the perturbation. In contrast, a system that responds to the perturbation in the opposite direction is called a *negative feedback* system. . . . The end result of a positive feedback is often amplifying and "explosive," i.e. a small perturbation results in big changes.

Another case of positive feedback in the Arctic is the melting of tundra permafrost (no longer so perma-), which releases vast quantities of the super-greenhouse gas methane from the rotting of formerly frozen vegetation, along with the evaporation of a weirdly flammable ice in the permafrost called methane hydrate. More methane in the atmosphere leads to more melting of permafrost, and so on. Oh, and also, with Arctic warming, the tree line rapidly moves north, so dark conifer forest replaces pale tundra, absorbing yet more solar heat, and another positive feedback gets going.

One important negative feedback may be operative, but its mechanisms are mysterious. Either because of atmospheric changes or human behavior, the world's land areas are absorbing more carbon dioxide than they're releasing lately. "The amazing truth is that on a global scale, photosynthesis is greater than decomposition and has been for decades," says atmospheric scientist Scott Denning. "Believe it or not, plant life is growing faster than it's dying. This means land is a net sink for carbon dioxide, rather than a net source." It might be simple carbon-dioxide fertilization—additional CO_2 stimulates plant growth; that's why it's used in commercial greenhouses. It might be that longer growing seasons in boreal regions are causing greater forest growth. On the other hand, forest fire suppression by humans could be the cause; or countless abandoned farms growing back as forest; or overgrazing by cattle, leading to woody shrublands replacing grasslands; or excess nitrogren from agriculture and automobiles, fertilizing additional wild growth. Until the "mysterious sink" for carbon is figured out, our climate models will remain frustratingly vague and unpredictive.

For hundreds of millions of years a "crazily jumping climate" has been the norm on Earth, says glaciologist Richard Alley. These days, apparently, we are returning to that jumpy norm, thanks to abruptness mechanisms like positive feedback, trigger events, and threshold effects, none of which are well incorporated into the climate models yet. It may take some breakthroughs in nonlinear mathematics before that can happen. A good book on the subject is Fred Pearce's *With Speed and Violence: Why Scientists Fear Tipping Points in Climate Change* (2007).

• There have been some cataclysmic trigger events in the past. A vast freshwater lake in North America suddenly emptied into the North Atlantic 12,800 years ago, and that was the Younger Dryas instant deep freeze. Another bizarre event occurred 55 million years ago, when a trillion tons of methane burped out of the oceans from thawing methane hydrates (also called clathrates) on the sea floor. The sudden temperature rise of 8°C (14.5°F) extinguished two thirds of oceanic species and was nearly as catastrophic on land as the dinosaur-killing asteroid 10 million years earlier. According to Fred Pearce's book, something between 1 trillion and 10 trillion tons of frozen methane clathrates lurk on the seabed now. Their potential sudden release is fondly known as the clathrate-gun hypothesis. David Archer, a climate modeler at the University of Chicago, has said, "The worst-case scenario is that global warming triggers a decade-long release of hundreds of gigatons of methane, the equivalent of ten times the current amount of greenhouse gas in the atmosphere. We'd be talking about mass extinction."

There's another potential trigger at the South Pole. The vast West Antarctic Ice Sheet, *fortunately*, is safely perched on land, held in place by the Ross Ice Shelf. *Unfortunately*, the Ross Ice Shelf is melting with surprising speed. If the West Antarctic Ice Sheet slides and melts into the ocean, sea levels will suddenly rise by 16 feet. (Higher, really, because the Greenland ice sheet is also melting).

Threshold effects are sneaky. Incremental change goes along and everything looks fine, and then before you know it, the system has shifted massively and irreversibly into another state. These decades the tropical rain forests are as busy as ever creating their own rain and reflective clouds, locking up lots of carbon, helping to slow global warming, apparently untroubled by it. At some point, though, a threshold is reached. Then in an unstoppable cascade the rain forests melt like Arctic ice, leaving savannah, scrub, and desert in their place. The carbon sink is gone, the reflective clouds are gone, a zillion species are gone, and we can't get them back. What is the fatal threshold for rain forests? Researchers Richard Betts and Peter Cox think it is just 3°C (5.4°F) above what we have now. The 2007 IPCC report predicts that the demise of the rain forests will be under way by 2050.

There are known and unknown thresholds in the ocean. At around 14°C (57.2°F) the surface stratifies, keeping cold-water nutrients out of reach of sunlight. Algae can't grow, and a whole swath of ocean goes dead, its carbon-fixing capacity crippled. A similar critical point involves the acidification of seawater by excess CO_2 in the air. With greater acidity, the carbonate-forming organisms, like reef coral, shellfish, and the teeming diatoms of the open ocean, are disabled, and their existing carbonate structures start to dissolve. The oceans, which now absorb a third of humanity's CO_2 emissions, flip from carbon sink to carbon source, from solution to problem. Where is that threshold? We don't know yet.

In his 2005 book, *Climate Crash: Abrupt Climate Change and What It Means for Our Future*, science writer John Cox summarizes the depth of uncertainty that explains why climate models have thus far been unable to predict the past or present with the kind of accuracy we want:

> The climate system is nonlinear, which means its output is not always proportional to its input—that, occasionally, unexpectedly, tiny changes in initial conditions provoke huge responses. It is chockablock with feedbacks, loops of self-perpetuating physical transactions, operating on their own timescales, that amplify or impede other processes. This constant cross talk of positive and negative feedbacks is said to be balanced, more or less, at various critical thresholds in the system. Forced across such a threshold, by whatever external or internal triggering mechanism, important variables begin gyrating or flickering, and the system suddenly lurches into a significantly different semistable mode of operation, a new equilibrium. All of these variables, all of these timescales, make for a system that is full of surprises. . . . Climate is a precariously balanced nonlinear system that lurches between very different states of coldness, dryness, wetness, and warmth.

Climate is so full of surprises, it might even surprise us with a hidden stability. Counting on that, though, would be like playing Russian roulette with all the chambers loaded but one.

● Some climate events are already having an impact on humans. Despite our suppression efforts, forest fires are increasing everywhere because, as one science writer puts it, "with global warming, we don't get fire; we get fire squared." Large fires in the drying forests and newly cleared peat bogs (such as those in Indonesia) dump vast quantities of CO_2 into the atmosphere, which further warms and dries land vegetation, making it ever more flammable. A 2007 megafire in southern Greece caused the government of the once-popular Costas Karamanlis to fall. Persistent drought in Australia led, in 2007, to a switch from a climate-denialist prime minister to one whose first official act was to ratify the Kyoto Protocol. His administration soon had its own megafires to deal with.

In Europe, studies show warmer temperatures are moving north at 25 miles a decade, whereas animals and plants are moving north at only 3.75 miles a decade. That's a formula for extinction. Olive and avocado trees now grow outdoors in London, at Kew Gardens. With the overfished ocean becoming warmer and more acidic, vast swarms of jellyfish are drifting north, killing whole fish farms in the Irish Sea. In Africa, warmth-loving mosquitoes are carrying malaria and dengue fever to higher elevations, and even bringing tropical diseases to southern Europe.

The glaciers of the Tibetan Plateau, which feed all the rivers of China, north India, and Southeast Asia with meltwater, are now vanishing. Three billion people depend on those rivers. In addition the billion in India live, as they say, "at the whim of the monsoon" for rain. The monsoon in turn lives at the whim of the El Niño cycle, which is being disrupted at the whim of the mid-Pacific trade winds, which are slackening due to ocean warming.

How human societies will respond to climate calamities remains to be seen. At Global Business Network, we've been studying the likely consequences of a growing frequency of extreme events such as the 2003 heat wave that killed thirty-five thousand in France and Italy. Nils Gilman at GBN notes that "while a single extreme event may be relatively easy to withstand, a second in succession is likely to be far more devastating, as normal resiliency measures are built to deal with one but not multiple consecutive extreme events." Governments, he concludes, "will experience climate change not as a smooth transformation, but rather as a series of radical discontinuities—as a series of bewildering 'oh shit' events. Environmentally failed states are a nontrivial possibility."

Repetition knocks you down; duration kills you. Complex societies can handle drought, but not multidecade drought. That's the historic civilization killer, says archaeologist Brian Fagan. It brought down the ancient empires of the Middle East and Central America. When the rains fail, agriculture fails, the cities convulse and empty, and what's left of the society builds shacks in the ruins of its former glory. In this century, the effects of rising sea levels, catastrophic as they may be, could look temporary and fixable compared to the effects of permanent drought.

• "We have to understand that the Earth system is now in positive feedback and is moving ineluctably toward the stable state of one of the past hot climates," atmospheric chemist James Lovelock told a Royal Society audience of scientists in 2007. "I can't stress too strongly the dangers inherent in systems in positive feedback."

Two of Lovelock's books, *The Revenge of Gaia* (2007) and *The Vanishing Face of Gaia* (2009), give the clearest warning yet of the extreme dangers we face and how radical our measures may have to be to deal with them. I've learned to trust Lovelock's judgment ever since 1974, when a magazine I edited, *CoEvolution Quarterly*, was the first to publish his Gaia hypothesis, coauthored with microbiologist Lynn Margulis. Since then, their idea of Earth as a self-regulating living system "comprised of physical, chemical, biological, and human components" steadily matured from hypothesis into theory; it became formalized as Earth Systems Science, and it has won Lovelock no end of prizes.

I phoned Jim Lovelock after his Royal Society talk to get details on why the gentle optimist I've known for three decades is so alarmed. "The year 2040 is when the IPCC is estimating that Europe, America, and China become uninhabitable for the growth of food," he said. "They're grossly underestimating the rate of temperature rise, so that 2040 may be 2025. People don't realize how little time we've got. The planet really is on the move."

"On the move toward what?" I asked.

He said: "I don't think there's much doubt at all now amongst those few of us that have worked on the problem, that the system is in the course of moving to its stable hot state, which is about 5 degrees Celsius globally higher than now. Once it gets there, negative feedback sets in again, and the whole thing stabilizes and regulates quite nicely. What happens is, during that period, the ocean ceases to have any influence on the system, or hardly any. It's run entirely by the land biota. That's what happened in the past, anyway. There's a good deal of geologic evidence; the best evidence comes from the 55-million-years-ago event. The Arctic ocean temperature was about 23° Celsius [73.4° F]—crocodiles swam around in it. The whole damn planet was tropical, probably. And will be again, if it goes on the way it's going. The equatorial regions were a hell of a lot drier than they are now. You see that already happening."

I asked him what might be the human carrying capacity in that hotter, stable Earth. "Oh, I think it's less than a billion," he said. "It will be too hot for things to grow." Then he added: "The earth will continue to move to its hot state almost regardless of what we do. Peter Cox at the Hadley Centre in our country has done some very careful analysis on how little CO_2 is needed to start the automatic jump from the cool to the hot state, and it's an astonishingly and worryingly small quantity. He probably doesn't want to be quoted. It turns out to be about a quarter of a gigaton of carbon per year. Now that compares with the eight gigatons that we're actually emitting to the atmosphere. So you'd have to cut back below that level to keep it stable, and you wouldn't succeed if it's already on course up towards its hot state. You're not going to turn it back."

● That's bleak. If the transition to a less livable Earth is already under way, we're ants on a burning log. We can rush around all we want; there's nothing in our ant repertoire that can fix the problem.

But we know a couple of things. We know the worst that can happen. We know that we probably have to extend our repertoire of capabilities to either head it off or live with it. The three broad strategies for dealing with climate change are *mitigation, adaptation,* and *amelioration*. Mitigation, cutting back on greenhouse gas emissions, has been called avoiding the unmanageable. Adaptation, then, is managing the unavoidable—moving coastal populations to higher ground, developing drought-tolerant agriculture, preparing for masses of climate refugees, and keeping resource warfare localized. And amelioration is adjusting the nature of the planet itself through large-scale geoengineering.

Civilization is at risk, but civilization is the problem. The key positive feedback in the current Earth system is us. Accelerating wealth (especially in developing countries these days), a still-growing human population, and accelerating industry are pouring overwhelming quantities of greenhouse gases into the atmosphere. As Australian biologist Tim Flannery puts it, "The metabolism of our economy is now on a collision course with the metabolism of our planet."

• If Lovelock's is the worst-case climate scenario—Earth stabilizes at 5°C (9°F) warmer; a fraction of the present human population survives— then what is the best case? What can we hope for? The person with the most realistic numbers is Saul Griffith, a materials scientist and inventor who received a MacArthur "genius" award in 2007. To begin with, he says, "It is not accurate to say, 'We can still stop climate change.' We are now working to stop worse climate change or much-worse-than-worse climate change."

The most common statement of an achievable goal for dealing with climate these days is leveling off at 450 parts per million (ppm) of CO_2 in the atmosphere, so Griffith builds his analysis around that outcome. We are currently at about 387 ppm and rising fast—each year it goes up more than 2 ppm. Griffith reminds everyone that the hope with the 450 ppm goal is that it will involve a global temperature rise of only about 2°C (3.6°F), and that is expected to mean "large loss of species, more severe storms, floods and droughts, refugees from sea level rise, and other unpalatable, expensive and inhumane consequences."

A convenient measure of energy generation is the gigawatt: a billion watts. A large coal-fired plant generates a gigawatt of electricity in a year; so does Hoover Dam; so does a nuclear reactor. Multiply that times a thousand, and you have the terawatt—a trillion watts. Humanity currently runs on about 16 terawatts of power, most of it from the burning of fossil fuels. It's like leaving 160 billion 100-watt lightbulbs on all the time. That's what is loading the atmosphere with lethal quantities of carbon dioxide.

Griffith calculates that, in order to keep the atmospheric concentration of CO_2 at no more than 450 ppm, humanity has to do something that is almost unimaginably difficult. We have to cut our fossil fuel use to around 3 terawatts a year, which means we have to produce all the rest of our power from non-fossil-fuel sources, *and* we have to do it in about twenty-five years or it will be too late to level off at 450 ppm.

So, Griffith says, "Imagine someone said you need 2 terawatts of wind, 2 terawatts of photovoltaic solar, 2 terawatts of solar thermal, 2 terawatts of geothermal, 2 terawatts of biofuels, and 3 terawatts of nuclear to give you 13 new clean terawatts. You add the existing 1.5 terawatts of biofuels and nuclear that we already use. You can also get 3 terawatts from coal and oil. That would give humanity around 17.5 terawatts—that allows for a little growth over the 16 terawatts we currently use. What would it take to do all that in 25 years?"

Here's the answer: "Two terawatts of photovoltaic would require installing 100 square meters of 15-percent-efficient solar cells every second, second after second, for the next 25 years. (That's about 1,200 square miles of solar cells a year, times 25 equals 30,000 square miles of photovoltaic cells.) Two terawatts of solar thermal? If it's 30 percent efficient all told, we'll need 50 square meters of highly reflective mirrors every second. (Some 600 square miles a year, times 25.) Two terawatts of biofuels? Something like 4 Olympic swimming pools of genetically engineered algae, installed every second. (About 61,000 square miles a year, times 25.) Two terawatts of wind? That's a 300-foot-diameter wind turbine every 5 minutes. (Install 105,000 turbines a year in good wind locations, times 25.) Two terawatts of geothermal? Build three 100-megawatt steam turbines every day—1,095 a year, times 25. Three terawatts of new nuclear? That's a 3-reactor, 3-gigawatt plant every week—52 a year, times 25."

Add it up, and when you're done, you've got an area about the size of Australia—"Call it Renewistan," says Griffith—covered with stuff dedicated to generating humanity's energy. That's not counting transmission lines, energy storage, materials, and support infrastructure, plus the costs of shutting down all the coal plants, oil refineries, etc. I asked Saul Griffith if he thinks we can really do it. "Technically," he said, "it is possible. Industrially, humanity has the collective capacity. But politically, I don't see how." Then he added, "But we have to try. Why else bother to be human and be in this game?"

✦

A tranquil climate, we're coming to realize, is one of the "ecosystem services" that civilization requires in order to prosper; indeed, to survive. The only nonjumpy period in all of climate history (apart from the vast frozen stillnesses of the nine major ice ages) is the relatively benign "long summer" of the past ten thousand years during which humans developed agriculture, cities, and complex societies. Of course we take a gentle climate for granted; civilization has never experienced anything else.

How do we value ecosystem services? The usual panoply (food, water, air, energy, drugs, decomposition, delight, and so on) defies economic valuation, but that doesn't stop people from trying. One ecology textbook puts the number at more than $40 trillion a year, close to the world's current gross domestic product. The hope seems to be that once we know how to value ecosystem services, we'll know how to manage ourselves in relation to them.

Once upon a time, I dreamed that economics would eventually swell up and include ecology, and we would no more be misled by notions of "externalities." Now I'm not so sure. I recall a friend leaning on me to admit that ecology and economics are the same thing. "No, damn it," I said. "Ecology is devoid of intention, and economics is made of little else." (I suspect that my friend was on to something, though, because economics enthusiasts and ecology enthusiasts share an affliction. Conservatives think that the self-organizing properties of a market economy are a miracle that must not be messed with. Greens think that the self-organizing properties of ecologies are a miracle that must not be messed with.)

In one of the most influential Green books, *Natural Capitalism* (1999), Paul Hawken and Amory Lovins propose replacing industrial capitalism, which "liquidates its [natural] capital and calls it income," with a natural capitalism based on higher efficiency in everything, biology-inspired industrial processes, a focus on services instead of products, and restoration of the all-sustaining envelope of natural systems. It's a good book with a helpful metaphor.

• But I find it more fruitful to think of ecosystem services as infrastructure. A bridge is infrastructure, and so is the river under it. Both support our life, and both require maintenance, which has to be paid for somehow. Radio spectrum is infrastructure, and so is an intact ozone layer. Both support our life, and both require international agreements to avert a "tragedy of the commons."

Between headlong industrial capitalism and a necessarily patient natural capitalism is a pace gap that is hard to bridge. With infrastructure, however, we already think in terms of duration and responsibility, so it's no stretch to extend that thinking to natural systems. When there are problems with built infrastructure, we're used to solving them with science, engineering, collaborative public agreements, and financial instruments such as bonds and public-private contracts. Those tools apply just as well to natural infrastructure.

Oddly enough, although humans have been building infrastructure for thousands of years, it's still an intellectual no-man's-land. I've yet to find any economic theory of infrastructure. One wry definition of infrastructure is: "something gray, behind a chain-link fence." The message is: "Don't look, don't touch, don't even think about what this gray thing is for." We're trained to overlook infrastructure.

There are some exceptions. People like the romanticism of railroads and admire bridges and ships. Small towns decorate their water towers. But working mines, containership ports, power plants, power lines, cellphone towers, refineries, dumps, sewerage—all bear one sign: KEEP OUT. Those places are left to the workers, who are low-status.

One might say exactly the same about ecosystem infrastructure, such as watersheds, wetlands, fisheries, soil, and climate. As the truism says, we only notice infrastructure when it doesn't work. And so, a deep bow of thanks is due to the environmentalists who for decades have been drawing attention to dangerous breakdowns of natural infrastructure and setting about the protection and restoration of watersheds, wetlands, fisheries, soil, climate, and the rest. Without their warnings and work, we would be in a far worse situation than we are.

• How did we start worrying about climate? In 1948 a conservationist named Fairfield Osborn wrote a book titled *Our Plundered Planet* (the first jeremiad of its kind) and, with Laurance Rockefeller, founded the Conservation Foundation in New York. In 1958 Charles Keeling began his epic project measuring the atmospheric concentration of CO_2. When the worrying upward trend of that concentration became apparent, Osborn's Conservation Foundation assembled the first climate change conference in 1963; this resulted in a paper, "Implications of Rising Carbon Dioxide Content of the Atmosphere." According to Spencer Weart's *Discovery of Global Warming* (2004), "Their report warned that the doubling of CO_2 projected for the next century could raise the world's temperature some 4°C (7.2°F), bringing serious coastal flooding and other damage." The Conservation Foundation urged renewed funding for Keeling's CO_2 project and pressed the National Academy of Sciences to pay attention to the subject. From then on, awareness of climate change ascended right along with the Keeling Curve. In 1971 Barry Commoner's environmentalist bestseller, *The Closing Circle*, gave an early public warning about greenhouse gases. In 1978 a young congressman from Tennessee, Albert Gore, held hearings on global warming, starring his Harvard teacher Roger Revelle, who had sponsored the Keeling CO_2 research.

After the OPEC oil embargo of 1973 focused the world's attention on energy, efficiency and renewability became core doctrine for environmentalists. Solar was hip. Wind-generated electricity began developing toward the mega-infrastructure it is now. Insulated windows were invented and refined. A by-product of all that innovation, especially from the drive toward efficiency, was that gigatons of carbon dioxide stayed out of the air. I was part of that, and you're welcome.

Unfortunately for the atmosphere, environmentalists helped stop carbon-free nuclear power cold in the 1970s and 1980s in the United States and Europe. (Except for France, which *fortunately* responded to the '73 oil crisis by building a power grid that was quickly 80 percent nuclear.)

Greens caused gigatons of carbon dioxide to enter the atmosphere from the coal and gas burning that went ahead instead of nuclear. I was part of that too, and I apologize.

• One more climate book to invoke here is *Plows, Plagues and Petroleum* (2005), by paleoclimatologist William Ruddiman. He examines the last 2.75 million years, dominated by dozens of ice ages, their period and amplitude driven by three intersecting astronomical cycles affecting solar intensity. Ice-core data from Greenland matches the cyclic theory closely until about five thousand years ago, when, in the midst of our current routine interglacial, the standard steep drop in atmospheric methane suddenly reversed and headed up. It's still going up. What the hell happened?

We happened. Ruddiman surmises that the cause was the sudden adoption of irrigation in China and South Asia for an agricultural innovation, wet rice cultivation. Vegetation rotted in the new artificial wetlands and emitted methane. As rice farming expanded, so did methane emission. Add the ever-growing numbers of methane-burping livestock, plus increasing forest burning for agriculture, and the anomaly is explained. Ruddiman wondered if a similar mysterious reversal and climb in atmospheric CO_2 eight thousand years ago might have a related explanation. It did. As human population took off with agriculture, forests were burned to make new fields and pasture. Whole societies grew and migrated, forests shrank, and the atmosphere became a greenhouse. On the old astronomical schedule, a new ice age should have begun a couple of thousand years ago. Ruddiman concludes, "A glaciation is now overdue, and we are the reason."

One further detail. What would explain the peculiar sudden dips in atmospheric CO_2 between 200 and 600, 1300 and 1400, and 1500 and 1750? Those dates happen to match major human diebacks from pandemics—Roman-era epidemics, the Black Death in Europe, and the devastation of North American native populations by European diseases. Each time, forests grew back rapidly over empty agricultural land and drew down carbon dioxide.

If Ruddiman is right, climate has been a human artifact, a highly sensitive one, for a long time. "The end of nature," to use Bill McKibben's famous book title, didn't begin two hundred years ago with the Industrial Revolution but ten thousand years ago with the agricultural revolution. Farm and pasture land now takes up over a third of the world's ice-free land surface. Ruddiman notes that "farming is not nature, but rather the largest alteration of Earth's surface from its natural state that humans have yet achieved." Furthermore, "A good case can be made that people in the Iron Age and even the late Stone Age had a much greater per-capita impact on the earth's landscape than the average modern-day person."

● Never mind terraforming Mars; we already live on a terraformed Earth. We've been inadvertently adroit at it for ten millennia, even heading off an ice age. *Unfortunately*, we're now excessively carbon-loading the atmosphere toward inferno, though *fortunately* some of the overheating has been masked by our other major air pollutant, particulate aerosols causing "global dimming." How much longer can we count on such a string of dumb good luck?

The terraforming thus far has been unintentional. Now that we have the curse and blessing of knowing what's going on, unintentional is no longer an option. "Nature" can't be counted on, having been compromised long ago. Gaia is no savior, since "she" likes ice ages and doesn't mind hot ages either. We're left with intention, with conscious design, with engineering. We finesse climate, or climate finesses us.

Of the tools that come to hand, this book will examine four that environmentalists have distrusted and now need to embrace, plus one we love that has to be scaled up. The unwelcome four are urbanization, nuclear power, biotechnology, and geoengineering. The familiar one is natural-system restoration, which may be better framed as megagardening—restoring Gaia's health at every scale from local soil to the whole atmosphere.

One more positive feedback to take account of in the overall climate system is the autocatalytic—self-accelerating—technologies that can be deployed against the self-accelerating problems of world industrialization and against the positive feedbacks in climate itself. Our management of future technology acceleration has to reverse the effects of past technology acceleration. (Stopping present technology where it is would lead to Lovelock's uninhabitable hot world.) The goal is for the climate-plus-humans system to settle down to a healthy, stable negative-feedback regime.

Not all technologies are autocatalytic: New discoveries don't make every technology advance faster. Progress in automobile technology and wind technology makes better cars and wind generators but not better tools for the engineering itself. The current autocatalytic technologies that goose themselves into exponential growth are infotech (including computers, communications, and artificial intelligence), biotech, and nanotech (which is blurring into biotech). What's more, they stimulate each other in a mutual catalysis that at times results in hyperexponential growth of power.

Forty years ago, I started the *Whole Earth Catalog* with the words, "We *are* as gods, and might as well get good at it." Those were innocent times. New situation, new motto: "We are as gods and *have* to get good at it." The *Whole Earth Catalog* encouraged individual power; *Whole Earth Discipline* is more about aggregate power.

The scale of the climate challenge is so vast that it cannot be met solely by grassroots groups and corporations, no matter how Green. The situation requires government fiat to set rules and enforce them. Specifically, the four major energy-using governments—the European Union, the United States, China, and India—have to get tough. If all four do the right thing, there's hope. So far the European governments have led the way.

Our civilization caused climate change, and now it is undertaking to cause climate nonchange. At the end of the exercise (if it's successful), climate will be the same but civilization probably won't. We will be more transformed by our efforts to stabilize climate than by anything else we do in this century. If we fail to stabilize climate, our civilization will either be gone or unrecognizable.

◆

Who wrote this book?

I turned seventy during the writing in 2008 and 2009. In seven decades, I've enjoyed the instruction of living downstream from a good many of my own and other people's mistakes. As the old joke goes: How do you build good judgment? (Experience!) How do you build experience? (Bad judgment!) Because I'm an ecologist by training, a futurist by profession, and a hacker (lazy engineer) at heart, my bent is scientific rigor, geoeconomic perspective, and an engineer's bias, which sees everything in terms of solvable design problems.

In keeping with professional forecaster practice, my opinions are strongly stated and loosely held—strongly stated so that clients can get at them to conjure with, loosely held so that facts and the persuasive arguments of others can get at them to change them. My opinion is not important; it's just a tool. The client's evolving opinion is what's important. Your evolving opinion is what's important. If you're reading this book just to reinforce your present opinions, you've hired the wrong consultant.

I'm a lifelong environmentalist. My voice piped at age ten: "I give my pledge as an American to save and faithfully to defend from waste the natural resources of my country—its air, soil, and minerals, its forests, waters, and wildlife." I got infected by that Conservation Pledge through the magazine *Outdoor Life* and proceeded to paste it on everything and everyone around me. Since the concept of *pledge* has long been rendered meaningless by the surreal Pledge of Allegiance that American schoolchildren have to recite, what I meant in 1948—and mean now—is: "I declare my intent to save and defend from waste the world's natural resources—its air, soil, and minerals, its forests, waters, and wildlife."

I graduated with a degree in biology from Stanford in 1960, having focused mainly on evolution and what was then the very low-status field of ecology. One of my teachers, the later-renowned Paul Ehrlich, encouraged me to publish the results from my only fieldwork, concerning two species of tarantulas out back of Stanford that appeared to be permanently mingled, in violation of Gause's principle, which states that no two species can long occupy the same niche together. Instead of publishing, I went off to the army to be an officer.

Some of the Green adventures I had after that will turn up later in these pages. My previous books (on new media and adaptive buildings) have been journalistic essays—reports by an outsider. This one is journalistic too, but it's written from inside its subject; that's what makes it a manifesto. Some of the issues I'm writing about I have a stake in; some of the people I'm writing about are friends. Where I think my personal experience has some relevance, I'll throw it in.

There are two things I won't attempt in this book. The goal of the environmental movement is to manage the commons well—meaning for everyone and for the long term. A great service would be to inventory and praise the countless environmental organizations and success stories that have kept the commons—air, forests, soil, oceans, animal life—as healthy as they are. I'm not doing that here. Another important service would be to inventory and condemn the innumerable cases in which governments, companies, and property owners have done their best to mismanage the commons for private and short-term gain, meanwhile disparaging and thwarting environmentalists. It would be fun and useful to compile a bestiary of such behaviors and examine their constituent pathologies, but I have other fish to fry.

Whole Earth Discipline carries on something that began in 1968, when I founded the *Whole Earth Catalog*. I stayed with the *Catalog* as editor and publisher until 1984, adding a magazine called *CoEvolution Quarterly* along the way. The Whole Earth publications were compendia of environmentalist tools and skills (along with much else) and explicitly purveyed a biological way of understanding. Peter Warshall wrote and reviewed about watersheds, soil, and ecology. Richard Nilsen and Rosemary Menninger covered organic farming and community gardens. J. Baldwin was an impeccable source on "appropriate technology"—solar, wind, insulation, bicycles. Lloyd Kahn wrote about handmade houses. We promoted bioregionalism, restoration, and "reinhabitation" of one's natural environment. There's now an insightful book about all that by Andrew Kirk—*Counterculture Green: The Whole Earth Catalog and American Environmentalism* (2007).

These days I divide my time between Global Business Network and an idiosyncratic foundation. In the 1990s, when inventor Danny Hillis came up with an idea to help people think long-term by building a monumental ten-thousand-year clock, I responded by cofounding The Long Now Foundation with him in 1996. "Fostering long-term responsibility" is its mission. The "long now" is defined as the last ten thousand years and the next ten thousand years. That is the reach of humanity's current decisions.

● Lovelock said, "The planet really is on the move." So is civilization, now completing a process it began ten millennia ago—moving to town. Ecopragmatism in this century has to begin with understanding what humanity's momentous transition from rural to urban is made of, and what it portends. The subject has so much news, I'll give it two chapters.

A "city planet" needs city power—grid electricity. At present, the best low-carbon source is nuclear. I'll explore a chapter's worth of how that fits into a climate-driven Green agenda and then do the same for genetic engineering for two chapters, because I believe biotech offers a major tool for reducing the overwhelming impact of agriculture on natural infrastructure, and new discoveries about genes and microbes are transforming the science of ecology.

Science has long informed the environmental movement. Now it must take the lead, because we are forced to enter an era of large-scale ecosystem engineering, and we have to know what the hell we're doing. That sermon gets a chapter. Beavers are benevolent ecosystem engineers; so are soil-enriching earthworms; so were American Indians, who terraformed a continent; so are all of us who work on restoring natural infrastructure. A chapter on that subject leads straight to the book's conclusion: our obligation to learn planet craft, to be as life-enhancing as any earthworm, in the big yard.

The footnotes for this chapter may be found online at
www.sbnotes.com. Live links to sources are included.

·2·
City Planet

Against the dark screen of night, Vimes had a vision of Ankh-Morpork. It wasn't a city, it was a *process*, a weight on the world that distorted the land for hundreds of miles around. People who'd never see it in their whole life nevertheless spent that life working for it. Thousands and thousands of green acres were part of it, forests were part of it. It drew in and consumed . . .

. . . and gave back the dung from its pens, and the soot from its chimneys, and steel, and saucepans, and all the tools by which its food was made. And also clothes, and fashions, and ideas, and interesting vices, songs, and knowledge, and something which, if looked at in the right light, was called civilization. That was what civilization *meant*. It meant the city.

—Terry Pratchett, *Night Watch*

Cities are wealth creators; they always have been. They are population sinks, and always have been. Just as agriculture raised the world's carrying capacity for humans, so did cities. The death rate from "constant battles" declined with urbanization, LeBlanc says, because "as the city folk make tools and improve the technologies that make the farming more efficient, more people can live in the city instead of farming, and the cities grow. To some degree, this process solves the resource/population pressure found among farmers."

The ten-thousand-year flow of people to cities has become a torrent. In 1800 the world was 3 percent urban; in 1900, 14 percent urban; in 2007, 50 percent urban. The world's population crossed that threshold—from a rural majority to an urban majority—at a sprint. We are now a city planet, and the Greener for it, as I'll show. But environmental projects in this century can prosper only if we understand what is really going on in cities and figure out how to work with it.

At the current rate, humanity may well be 80 percent urban by midcentury. Every week there are 1.3 million new people in cities. That's 70 million a year, decade after decade. It is the largest movement of people in history.

What is really going on?

"In the village, all there is for a woman is to obey her husband and relatives, pound millet, and sing. If she moves to town, she can get a job, start a business, and get education for her children." That remark at a conference in 2001 exploded my Gandhiesque romanticism about villages. The speaker was Kavita Ramdas, head of the Global Fund for Women. Ever since she spun my head, I've been asking travelers back from remote places what they noticed out in the countryside. Their universal report: The villages of the world are emptying out, everywhere.

Demographers talk about "push" and "pull" motivations driving the migration to cities. Push feels like this: Life in your village is dull, backbreaking, impoverished, restricted, exposed, dangerous, and static. Brigands get you, an accident gets you, disease gets you, and there's no help nearby. You work like hell; then the weather changes, and you don't have crops to eat or sell. Visit your relative in town and you see what "pull" means. In the city, life is exciting; work is less grueling; you're far better paid; you're free to move around and change jobs; you have some privacy; you're less vulnerable; and you have upward mobility. Will you put up with slum conditions for all that? In a heartbeat. "City air makes you free," said the Renaissance Germans. History may view the urban release of the European Renaissance as mild compared to what's going on now.

The move to town is a liberation. A *New York Times* story in 2005 related that

> Gandhi idealized villages as the way to return Indians to their
> precolonial state. B. R. Ambedkar, the Dalit, or untouchable,
> leader who helped write India's Constitution, saw it differently:
> he called villages a cesspool, "a den of ignorance, narrow-

mindedness and communalism," and urged untouchables to flee them for urban anonymity.

The same article gave an example of the strongest reason to migrate from India's 600,000 villages to cities such as Surat, with a population of 3.5 million:

> Rajesh Kumar Raghavji Santoki, 28, had tried farming for a year at home, and given up in the face of a water shortage. After just three years in Surat, he was earning in a month more than the $500 his farmer father earned in a year. He owned a house, a motorcycle and a van.

Multiply his motivation by 900 million—the 70 percent of India's 1.3 billion still living in rural areas. Multiply it by 2.8 billion, the number of people still rural throughout the developing world. At the same time that opportunity in the cities is becoming more attractive, in many places the countryside keeps getting rougher. The land is depleted by overuse, landholdings shrink as they are divided among successive generations, and civil strife is a frequent threat. Many of my contemporaries in the developed world regard subsistence farming as soulful and organic, but it is a poverty trap and an environmental disaster. When subsistence farms are abandoned, the trees and shrubs, no longer gathered for firewood, quickly return, and so do the wild animals no longer hunted and trapped for bush meat.

• In developed regions like North America and Europe, the rural push and urban pull are different but just as compelling. Lou Reed sings, "When you're growing up in a small town / You know you'll grow down in a small town./ There's only one good use for a small town: / You hate it, and you know you'll have to leave." In America's northern high plains, cities like Fargo, Bismarck, and Grand Forks are thriving, but the rest of the vast grasslands is emptying, leaving ghost towns and decaying farmhouses. *National Geographic* describes "a sense of things ebbing, of churches being abandoned, schools shutting down, towns becoming ruins." Some megafauna—moose and mountain lions—are coming back. The whole high-plains region from eastern Montana to northern Texas is headed toward becoming the "buffalo commons" that environmentalists long for.

In developed countries like the United States the migration is from boring, lonely, and hard to exciting, busy, and pleasant—toward coasts, sun, and densely citified regions called megapolitan areas, such as the one encompassing everything on the eastern seaboard from Boston to Richmond, Virginia, or the one I live in, reaching from San Francisco to Reno, Nevada—"from sea to ski." All over the developed world, once-thriving remote fishing villages are emptying. The fishing, alas, carries on more intensely than ever, but now in urban-based factory ships. When Communism collapsed, the formerly subsidized small towns of central Russia and Eastern Europe instantly lost their young people and their future.

But the main event is not in Europe or North America. The United States has 49 cities with populations over a million; China has 160. Since 1950 in China, 300 *million* people have moved to the cities, and *another* 300 million are expected in the next few decades. That's nearly half of China's total population of 1.3 billion on the move. It is typical of the developing world.

According to some historians, "Civilization is what happens in cities." (*Civilization* and *city* are in the same Latinate word group, along with *civil*, *civics*, *citizen*, etc.) I live in California, but I avidly read the *New Yorker* and the *New York Times*, just as the French read *Paris Match* and the English read the London newspapers. You can characterize any nation by its largest city, and study the progress of any era by examining the largest cities of its time. Most of the people I know think of the world in terms of London, New York, Paris, Berlin—the great Western metropolises. Those were indeed the largest cities a hundred years ago. In 1900 London had a population of 6.5 million; New York had 4.2 million; followed by Paris, Berlin, Chicago, Vienna, Tokyo, Saint Petersburg, Manchester, and Philadelphia. Tokyo is the only surprise in that top-ten list.

Fifty years later, in 1950, the leading ten cities had doubled in size, and Shanghai, Buenos Aires, and Calcutta had joined the list. Fifty years after that, in 2003, the top ten cities had further tripled in size, but that was the least of their changes. The leaders now were Tokyo with 35 million, Mexico City with 19 million, New York still in the game with 18 million, São Paulo with 18 million, Mumbai with 17 million, Delhi with 14 million, Calcutta with 13 million, Buenos Aires with 13 million, Shanghai with 13 million, and Jakarta with 12 million. Those big numbers are big events. By 2015, according to United Nations predictions, the top-ten roster will be joined by Dhaka, in Bangladesh, and Lagos, in Nigeria; and coming on fast will be Karachi, Cairo, Manila, Istanbul, Lima, Tehran, and Beijing.

The trend is pretty clear. The "rise of the West" is over. The world looks the way it did a thousand years ago, when the ten largest cities were Córdoba, in Spain; Kaifeng, in China; Constantinople; Angkor, in Cambodia; Kyoto; Cairo; Baghdad; Nishapur, in Iran; Al-Hasa, in Saudi Arabia; and Patan, in India. As Swedish statistician Hans Rosling says, "The world will be normal again; it will be an Asian world, as it always was except for these last thousand years. They are working like hell to make that happen, whereas we are consuming like hell."

● It may be distracting, though, to focus just on the world's twenty-four megacities—those with a population over 10 million. The real action is in what the United Nations calls small cities (fewer than 500,000 inhabitants; home to half of the world's city dwellers) and intermediate cities (1 million to 5 million, where 22 percent of urbanites live). A UN report points out: "They are often the first places where the social urban transformation of families and individuals occurs; by offering economic linkages between rural and urban environments, they can provide a 'first step' out of poverty for impoverished rural populations and a gateway to opportunities in larger cities."

The Marxist scholar Mike Davis gives perspective on the phenomenon in his 2006 book, *Planet of Slums*:

> In Africa . . . the supernova-like growth of a few giant cities like Lagos (from 300,000 in 1950 to 10 million today) has been matched by the transformation of several dozen small towns and oases like Ouagadougou, Nouakchott, Douala, Antananarivo

and Bamako into cities larger than San Francisco or Manchester. In Latin America, where primary cities long monopolized growth, secondary cities like Tijuana, Curitiba, Temuco, Salvador and Belém are now booming.

In other words, more and more news will be coming from cities most people in the West have never heard of. Developing countries are urbanizing at a rate and volume qualitatively different from what happened in Europe and North America—three times faster and nine times bigger. Beyond our horizon of attention, the world is being transformed.

● Of all human organizations, cities are the longest-lived. The oldest surviving corporations, Stora Enso in Sweden and the Sumitomo Group in Japan, are about 700 and 400 years old, respectively. The oldest universities, in Bologna and Paris, have lasted only 1,000 years so far. The oldest living mainstream religions, Hinduism and Judaism, date back about 3,500 years. But the town of Jericho has been continuously occupied for 10,500 years. Its neighbor Jerusalem has been an important city for 5,000 years, even though it was conquered or destroyed thirty-six times and endured eleven conversions from one religion to another. Many cities die or decline to irrelevance, but some thrive for millennia.

I suspect that one cause of their durability is that cities are the most constantly changing of organizations. In Europe they consume 2 to 3 percent per year of their material fabric (buildings, roads, and other construction) through demolition and rebuilding. Effectively, a whole new city takes shape every fifty years. In the United States and the developing world, the turnover is even faster. Yet despite all the physical metamorphosis, something about a city remains deeply constant. Some combination of geography, economics, and cultural continuity ensures that even a city destroyed by war (Warsaw, Tokyo) or fire (London, San Francisco) will often be rebuilt and retain its identity.

● Cities are horrendously expensive, both environmentally and economically, but they more than earn their keep. "Cities make countries rich.

Countries that are highly urbanized have higher incomes, more stable economies, stronger institutions. They are better able to withstand the volatility of the global economy than those with less urbanized populations." So notes the United Nations Human Settlements Programme (UN-HABITAT), which was impelled to its city-boosting position by revelations in the worldwide data it has been gathering since 1978.

The reversal of opinion about fast-growing cities—from bad news to good news—began with *The Challenge of Slums*, a 2003 UN-HABITAT report. The book's reluctant optimism came from its groundbreaking fieldwork—thirty-seven case studies in slums worldwide. Instead of just compiling numbers and filtering them through remotely conceived theories, the researchers hung out in the slums, talking to people. They came back with an unexpected observation: "Cities are so much more successful in promoting new forms of income generation, and it is so much cheaper to provide services in urban areas, that some experts have actually suggested that the only realistic poverty reduction strategy is to get as many people as possible to move to the city."

In 2007 the United Nations Population Fund gave that year's report the upbeat title *Unleashing the Potential of Urban Growth*. The lead author, Canadian demographer George Martine, wrote, "Cities concentrate poverty, but they also represent the best hope of escaping it." He declared in a talk that "80 to 90 percent of GNP growth occurs in cities" and that "the half of the world's population living in cities occupies only 2.8 percent of the world's land area." He went on to say, "In cities, concentration and density make it easier to provide social services. Education, health, sanitation, water, electrical power—everything is so much easier and cheaper on a per capita basis."

Cities have always benefited from what are called "economies of agglomeration"—density accelerates economic activity—and lately they have gotten a further boost from globalization. Telecommunications and markets bypass national borders ever more easily. In some developing countries where the national government has been discredited, everybody just works around it. Aid organizations go straight to the cities, where the need is; and multinational corporations go straight to where the workers and emerging markets are, in the cities.

"The world's forty largest megaregions, which are home to some 18 percent of the world's population," writes urban theorist Richard Florida, "produce two-thirds of global economic output and nearly 9 in 10 new patented innovations." Whereas nations are defined by their boundaries, cities are densely connected nodes, making every city a world city to some degree, with the accompanying multipliers of cultural diversity, financial flow, and population flow. In the vast worldwide migration toward jobs, the poor hardly limit their travels to cities in their own countries. Also there is coming to be a large population of global urbanists who live at large in multiple cities. (In airports, they would say.) These peripatetics include hundreds of thousands of professionals and many of the world's wealthy—there are over 10 million millionaires worldwide. As cities intensified globalization, and globalization enriched cities, world trade grew elevenfold between 1980 and 2004—from $580 billion to over $6 trillion. (That growth rate hit a wall in 2008 with the world financial crisis.) Growing affluence has brought a return of the city-state in some places. The independent power of a Singapore or Dubai rivals that of ancient Athens or fifteenth-century Venice.

• A new theory is upsetting our idea of what cities are and can become. Through a phenomenon known as Kleiber's law, organisms become more metabolically efficient as they scale up—from shrew to elephant, say. Cities do the same. "One of the basic principles of cities is that it's more efficient to bring people together," says physicist Geoffrey West. "You need a little bit less of everything per person. It's the exact same way in biology. As animals get bigger, they require less energy to support each unit of tissue." But organisms move more slowly as they increase in size (compare a shrew's whirring heart rate to the stately thump of an elephant's heart), whereas cities speed up as they get bigger.

That was the news in a landmark paper, "Growth, Innovation, Scaling, and the Pace of Life in Cities," which appeared in the *Proceedings of the National Academy of Sciences* in 2007; Geoffrey West was a coauthor. Looking at everything from patents to personal income to electrical cable length in a variety of cities, the researchers found that not only do cities increase their creativity with increasing size, but the relation is "superlinear": when a city doubles in size, it more than doubles its rate of innovation. A summary of the paper in the *Santa Fe Institute Bulletin* reported that

> Individual productivity rises (15% per person when the city doubles) as people get busier. Average walking speeds increase. Businesses, public spaces, nightclubs, and public squares consume more electricity. The city draws in more inventors, artists, researchers, and financiers. Wealth increases, as does the cost of housing.

City growth creates problems, and then city innovation speeds up to solve them. "Not only does the pace of life increase with city size," the authors wrote, "but so also must the rate at which new major adaptations and innovations need to be introduced to sustain the city."

The paper concluded, "We have shown that growth driven by innovation implies, in principle, no limit to the size of a city, providing a quantitative argument against classical ideas in urban economics." In other words, West told me, "Cities can go on growing forever. Look at the invention of the steam engine, the car, the digital revolution. What these advances all have in common is that they allowed cities to continue growing." If cities are concentrators of efficiency and innovation, an article about the scaling paper in *Conservation* magazine surmised, then, "the secret to creating a more environmentally sustainable society is making our cities bigger. We need more metropolises." (I am a contributing editor to *Conservation*.)

In Peter Ackroyd's *London: The Biography* (2000), he quotes William Blake—"Without Contraries is no progression"—and ventures that Blake came to that view from his immersion in London. "Wherever you go in the city," Ackroyd observes, "you are continually being assaulted by difference, and it could be surmised that the city is simply made up of contrasts; it is the sum of its differences." What drives a city's innovation engine, then—and thus its wealth engine—is its multitude of contrasts. The more and greater the contrasts, and the more they are marbled together, the better. The most productive city is one with many cultures, many languages, many neighborhoods, and more kinds of urban experience available than any citizen can keep track of. In this formulation, it is the throwing together of great wealth and great poverty in the urban stew that is part of the cure for poverty.

◆

The common theory of the origin of cities states that they resulted from the invention of agriculture: Surplus food freed people to become specialists. You can't have full-time cobblers, blacksmiths, and bureaucrats, the theory goes, without farms to feed them. Jane Jacobs upended that supposition in *The Economy of Cities* (1969). "Rural economies, including agricultural work," she wrote, "are directly built upon city economies and city work." It was so in the beginning, she argued, and continues to this day. Most farming innovations, for example, are city-based. When Rome collapsed, European agriculture collapsed. When crop rotation was reinvented in the twelfth century, it began around European cities and took two centuries to reach remote farms. In the eighteenth century, the revolutionary use of fodder crops like alfalfa to fix nitrogen in the soil was developed first in city gardens. American agriculture soared in the 1920s when hybrid corn was invented, not on a farm but in a New Haven, Connecticut, laboratory.

If agriculture didn't create cities, what did? Jane Jacobs thought it was trade. My guess, based on the "constant battles" view of history, is defense. The first urban invention, I'll bet, was a defendable wall, followed by rectangular buildings that allowed close packing of maximum residents within a minimum amount of wall. (Pastoral and hunter-gatherer buildings—yurts, tipis, hogans, wikiups, bomas, and the like—are round.) Just like the most ancient town dwellers of Mesopotamia, the agricultural Pueblo tribes of the American Southwest lived in dense fortresses several stories high, with no openings in the outer walls. Entry was by retractable ladders. When defense against raids by nomadic Apaches and Navajos became irrelevant after the conquest by whites, the Pueblos all dispersed into scattered buildings (except where high-rise density is maintained partly for tourists, as at Taos and Acoma). "The earliest meaning of 'town,' said the urban scholar Lewis Mumford, "is an enclosed or fortified place."

Agriculture, it appears, was an early invention by the dwellers of walled towns to allow their settlements to keep growing, as in Geoffrey West's formulation. Today's megacities rely on the same flow of innovation. A 2004 UN-HABITAT report proposed that

> Cities are engines of rural development. . . . Improved infrastructure between rural areas and cities increases rural productivity and enhances rural residents' access to education, health care, markets, credit, information and other services. On the other hand, enhanced urban-rural linkages benefit cities through increased rural demand for urban goods and services and added value derived from agricultural produce.

Nothing saves a village like a good road to town and a good cellphone connection.

When urban migration leaves fewer people on the land, the ones remaining can shift from subsistence farming on marginal land to more concentrated cash-crop agriculture on prime land. That's better for the city, better for the locals, and better for natural systems in the area. Aquifers recover; forests recover. A study in Panama showed what happened when people abandoned slash-and-burn agriculture to move to town: "With people gone, secondary forest has regenerated. Crucially, if protected from hunters, nearly every bird and mammal species found in primary forest has also been found in secondary." Fifty-five times more tropical rain forest is growing back each year than is being cut, according to a 2009 report on world forests from the UN: 38 million acres of primary forest is cut, but 2.1 billion acres of secondary forest is growing back on land that was once farmed, logged, or burned.

Yet another urban innovation is the environmentalist idea of protecting, preserving, and restoring natural systems. As societies become more urban, they become Greener in their sensibilities. As their cities become more globally oriented they pick up environmentalist ideas and practices— and demands—from abroad. All of that can, if encouraged, contribute to increasing protection for the countryside emptying of people and refilling with biodiversity.

✦

Peasants who leave the land take rural skills and values to the city slums with them. Building their own shelter is what they've always done, at a minuscule fraction of the cost of city-provided housing. Collaborating with extended family and neighbors in close proximity is nothing new to them, and neither is doing without elaborate infrastructure. Those are all the abilities they need to build the most creative urban phenomenon of our time, the squatter cities—the teeming slums of the uninvited that house a billion people now, two billion soon.

Let no one romanticize what the slum conditions are. New squatter cities typically look like human cesspools and often smell like them. Usually there are no facilities at all for sanitation, for water, for electricity, for transportation. Everyone lives in dilapidated shacks that are jammed together wall to wall, every room full of people. A typical squatter city, which may stretch for miles, has grown without a plan or government, in an area generally deemed uninhabitable: a swamp, a floodplain, a steep hillside, or a municipal dump; clustered in the path of a highway project or squashed up against a busy railroad line.

But the squatter cities are *vibrant*. Their narrow lanes are bustling markets, with food stalls, bars, cafés, hair salons, dentists, churches, schools, health clubs, and mini-shops trading in cellphones, tools, trinkets, clothes, electronic gadgets, and bootleg videos and music. This is urban life at its most intense. It is social capital at its richest, because everybody in a slum neighborhood knows everybody else intimately, whether they want to or not. What you see up close is not a despondent populace crushed by poverty but a lot of people busy getting out of poverty as fast as they can.

Perhaps the most extreme case is Mumbai, with 17 million people more densely packed than anywhere else in the world. The city is half slum, yet it generates one sixth of India's gross domestic product. Suketa Mehta, author of *Maximum City* (2004), wrote in 2007:

Why would anyone leave a brick house in the village with its two mango trees and its view of small hills in the East to come here? So that someday the eldest son can buy two rooms in Mira Road, at the northern edges of the city. And the younger one can move beyond that, to New Jersey. Discomfort is an investment. . . .

One brother works and supports the others, and he gains satisfaction from the fact that his nephew takes an interest in computers and will probably go on to America. Mumbai functions on such invisible networks of assistance. In a Mumbai slum, there is no individual, only the organism. There are circles of fealty and duty within the organism, but the smallest circle is the family. There is no circle around the self.

It's a place where your caste doesn't matter, where a woman can dine alone at a restaurant without harassment, and where you can marry the person of your choice. For the young person in an Indian village, the call of Mumbai isn't just about money. It's also about freedom.

By 2004 I knew something important was up with the rampant urbanization of the developing world, but I couldn't find much in the way of ground truth about it until the publication of *Shadow Cities: A Billion Squatters, a New Urban World*, by journalist Robert Neuwirth. His research strategy was to learn the relevant language and then live for months as a slum resident—in Rocinha (one of seven hundred *favelas* in Rio de Janeiro), in Kibera (a squatter city of 1 million outside Nairobi), in the Sanjay Gandhi Nagar neighborhood of Mumbai, and in Sultanbeyli, a now fully developed squatter city of 300,000 with a seven-story city hall, outside Istanbul. In each seemingly scary shantytown, Neuwirth found he could just walk in, ask around, find a place to rent, and start making friends. In Kibera he was the only white person for miles, and no one cared. He was frightened just once, when city police in Rio threatened him, apparently because he had neglected to bribe them.

Contrary to a standard assumption, Neuwirth discovered that the wretched quality of housing in squatter cities is never the main concern of the inhabitants. The sad fact is that when governments and idealistic architects try to help by providing public housing, those buildings invariably turn into the worst part of the slum. The people who build the shanties take pride in them and are always working to improve them. The real issues for the squatters, Neuwirth found, are location—they want to be close to work—and what the UN calls security of tenure: They need to know that their homes and community won't be suddenly bulldozed out of existence.

They don't worry about unemployment: Everyone works, including the children. They don't worry about telephone service: Everyone has a cellphone or access to one. Medical care is available, and so is food; famine is now a rural phenomenon. The greatest need in every squatter city is infrastructure—water, electricity, and sanitation. Not always the hotbeds of criminal activity that everyone assumed, some squatter communities are victimized by criminals from outside because they have no police protection. Though the squatters join forces for what the UN researchers describe as "cultural movements and levels of solidarity unknown in leafy suburbs," they are seldom politically active beyond defending their own community interests.

A depiction of contemporary slum reality even more vivid than Neuwirth's is an autobiographical novel by an escaped Australian prisoner who went into hiding in Mumbai's slums, joined its organized crime world, and fell in love with the city. The author of *Shantaram* (2005), Gregory David Roberts, writes with all of the intensity and journalistic detail of a Victor Hugo, but from a level of experience that Hugo never had.

Roberts found life in the slum a perpetual melodrama of relationships, dangers, quests, emotions, services, and confrontations so overpowering that he couldn't wait to leave; but when he did leave, he kept going back because he missed all that drama. Rob Neuwirth found a similar phenomenon—some people who earned enough money to leave the *favela* and get an apartment elsewhere found they were bored and lonely. They returned for the excitement and sense of community. In Mumbai, some slum dwellers who have been moved into free apartments want out. "Before," said one man, "there would always be four guys around your shanty. We sat, we chatted. Now it's like being caged in a poultry farm."

(Women in the apartments, however, cherish having their own toilet and privacy.)

Social cohesiveness is the crucial factor differentiating "slums of hope" from "slums of despair." This is where CBOs (community-based organizations) and the NGOs (national and global nongovernmental organizations) shine. Typical CBOs include, according to UN-HABITAT's *The Challenge of Slums* (2003), "community theatre and leisure groups; sports groups; residents associations or societies; savings and credit groups; child care groups; minority support groups; clubs; advocacy groups; and more. . . . CBOs as interest associations have filled an institutional vacuum, providing basic services such as communal kitchens, milk for children, income-earning schemes and cooperatives."

Robert Kaplan reported in 2007 that the NGOs in Bangladesh "represent a whole new organizational life-form." In the cities and villages, they "make an end run around dysfunctional governments," and "because Bangladeshi NGOs are supported by international donors, they have been indoctrinated with international norms to an extent unmatched by the private sector."

● Women play a pivotal role in all this. The UN report notes that CBOs "are frequently run and controlled by impoverished women and are usually based on self-help principles, though they may receive assistance from NGOs, churches and political parties." A major impact of the move from countryside to city is that it unleashes woman power. Lenders have learned that microfinance credit works best when provided to women instead of men, and women are the more responsible holders of property deeds. *The Challenge of Slums* summarizes: "In many cases, women are taking the lead in devising survival strategies that are, effectively, the governance structures of the developing world when formal structures have failed them. However, one out of every four countries in the developing world has a constitution or national laws that contain impediments to women owning land and taking mortgages in their own names."

It is so important to free up newly urbanized women from their traditional role as fetchers of water and fuel that, as the UN report drily suggests, "the provision of water standpipes may be far more effective in enabling women to undertake income-earning activities than the provision of skills training."

A 2007 UN report, *Growing Up Urban*, tells the story of Shimu, a woman in her early twenties who moved from a small village in northern Bangladesh to a garment-making job in Dhaka:

> For the first time in her life she had got rid of her husband, her in-laws, her village and their burdens. A few months after she arrived, Shimu, now able to support her children, mustered the courage to return to her town and file for divorce. . . .
>
> Shimu prefers living in Dhaka because "it is safer, and here I can earn a living, live and think my own way," she says. In her village none of this would have been possible. But she thinks that when she is older she will go back there. She plans to buy a piece of land and settle there.

That bond back to the village appears to be universal. *The Challenge of Slums* observes that "Kenyan urban folk who have lived in downtown Nairobi all of their lives, if asked where they come from, will say from Nyeri or Kiambu or Eldoret, even if they have never been to these places. They will be taken there to be buried on ancestral land when they die." The persistence of that soul-bond to the land could be a great asset for assuring eventual environmental recovery in the developing countries, when love of the land plays out as protection of the land.

Religious groups have a stronger support role in the slums than most people realize. As Mike Davis wrote in *Planet of Slums*,

> Populist Islam and Pentecostal Christianity (and in Bombay, the cult of Shivaji) occupy a social space analogous to that of early-twentieth-century socialism and anarchism. In Morocco, for instance, where half a million rural emigrants are absorbed into the teeming cities every year, and where half the population is under 25, Islamicist movements like "Justice and Welfare," founded by Sheik Abdessalam Yassin, have become the real

governments of the slums: organizing night schools, providing legal aid to victims of state abuse, buying medicine for the
sick, subsidizing pilgrimages and paying for funerals. . . . Pen-
tecostalism is . . . the first major world religion to have grown
up almost entirely in the soil of the modern urban slum. . . .
Since 1970, and largely because of its appeal to slum women
and its reputation for being color-blind, [Pentacostalism] has
been growing into what is arguably the largest self-organized
movement of urban poor people on the planet.

In the 2007 UN report, George Martine noted that "Rapid urbanization
was expected to mean the triumph of rationality, secular values and the
demystification of the world. . . . Instead . . . the growth of new religious
movements is primarily an urban phenomenon. . . . In China, where cities are growing at a breakneck pace, religious movements are fast gaining
adherents."

● To me the most compelling image of hope in squatter communities
is something you see everywhere—masonry and concrete building walls
with rebar sticking out the top, ready for further construction. On the
upper floors of hand-built high-rises, the rebar is there in the expectation
that eventually another story will be added to the building—space for a
related family or another source of rent. All around urban Turkey, you see
heaps of tile bricks in people's yards. When they get a little money, they
buy some bricks, which are impervious to currency inflation. When they
get some more money, they build a wall or two. Unfinished masonry holds
up fine against weather.
    In new squatter communities, and in ones that are constantly threat-
ened by demolition, the shack materials are cardboard, cloth, plastic, scrap
wood, flattened oil drums, and—the most prized—corrugated steel sheets.
Rob Neuwirth chants in *Shadow Cities:*

> Praise be to plastic pipe. All honor the prefab window. Bow down
> to sheets of old plywood, stock-model sinks, mass-produced

tile. Three cheers for cement and cinderblock. Exalt the lowly rebar. Let's hear it for quick-drying concrete. Hooray for easy plastic wiring, easy plug outlets, and modular telephone service.

Over time, the walls get solider and higher, the materials more durable. The magic of squatter cities is that they are improved steadily and gradually, increment by increment, by the people living there. Each home is built that way, and so is the whole community. To a planner's eye, squatter cities look chaotic. To my biologist's eye, they look organic.

Prince Charles has the same opinion. After visiting Mumbai's Dharavi slum, he told an audience in London, "I find an underlying, intuitive grammar of design that subconsciously produces [a place] that is walkable, mixed-use, and adapted to local climate and materials—which is totally absent from the faceless slab blocks that are still being built around the world to warehouse the poor."

According to urban researchers, squatters are now the predominant builders of cities in the world.

● Inside the homes of the older squatter communities is another surprise. Field researchers in Cambodia for the 2003 UN report found that

> All slum households in Bangkok have a colour television. The average number of TVs per household is 1.6. . . . Almost all of them have a refrigerator. Two-thirds of the households have a CD player, a washing machine, and 1.5 cellphones. Half of them have a home telephone, a video player and a motorcycle.

Back in 1970, Janice Perlman interviewed 750 residents in the *favelas* of Rio. Her resulting book, *The Myth of Marginality* (1976), observed that the *favelados* "have the aspirations of the bourgeoisie, the perseverance of pioneers, and the values of patriots." Thirty years later, in 2001, she went back to interview her original informants and their children. The changes were dramatic. While the residents of the *favelas* still suffer discrimination because of where they live, their literacy rate has gone from 5 percent among the original migrants to 94 percent in their children. Everyone now lives in brick buildings, with electricity, water, and indoor bathrooms. All have refrigerators, TVs, cellphones, and washing machines, and are more likely to have microwave ovens and computers than are middle-class people elsewhere in Rio. Two thirds had left the *favelas* for more legitimate neighborhoods, but many who stayed now have genuinely luxurious wall tiling and furniture sets in their homes.

A reporter from the *Economist* who visited Mumbai wrote that "Dharavi, which is allegedly Asia's biggest slum, is vibrantly and triumphantly alive. . . . In fluorescent-strip-lit shops, in snatched exchanges in the pedestrian crush, as a hookah is passed around a tea-stall, again and again, the stories are the same. Everyone is working hard and everyone is moving up."

✦

Slums are the scene of a world-changing economic event, but it escapes notice because it's designed to escape notice. Squatters don't formally own land or property. They don't pay taxes. They take no part in any permit or licensing process. They pay no attention to government-approved exchange rates. And yet they thrive economically, charging each other rent for space in buildings with no legal ownership, employing each other in unlicensed businesses, and selling each other all manner of services and goods—some of the goods pirated, some of the services criminal. This is what is called the "informal economy." It is to economic theory what dark energy is to astrophysical theory. It's not supposed to exist, but there it is, and it's huge.

While the informal economy specializes in being invisible to the formal world, of course it is highly visible to itself. This is where the social capital of a dense community pays off. Without formal property title, everyone knows who effectively owns a building and may charge rent for a room in it. If you have a skill as a language teacher or identity card forger or whatever, there's no need to advertise: Your customers will find you, and the officials won't.

The Challenge of Slums—the 2003 United Nations report—estimates that 60 percent of urban employment in the developing world is in the informal sector and that the informal economy has essential links to the success of the formal sector:

> The screen-printer who provides laundry bags to hotels, the charcoal burner who wheels his cycle up to the copper smelter and delivers sacks of charcoal . . . , the home-based crèche to which the managing director delivers her child each morning, the informal builder who adds a security wall around the home of the government minister, all indicate the complex networks of linkages between informal and formal.

In many cities, the roughest slums are pressed right up against the most affluent neighborhoods. It looks like grotesque inequity, and it is, but it's mainly an efficient economic event typical of city density—service supply and service demand cleaving close to one another. The maids, nannies, gardeners, and security guards walk to work.

The poor have time but no money, and the rich have money but no time; and so they deal. I find what needy people do with surplus time more interesting than what un-needy people do with surplus money. Ingenuity is the norm in the informal economy. For instance, there is a whole urban-farming subeconomy in the slums: Families save money and improve nutrition by growing their own food, and they can sell the produce a short distance from where it's grown. In the Medellín slums in Colombia, "people raise pigs on the third-floor roofs and grow vegetables in cut-open bleach bottles they hang from their window sills," according to blogger Ethan Zuckerman.

At the entry level, the informal economy is organized around pittances. Robert Kaplan wrote in 2008 that

> For the many rural newcomers to Bangladesh's cities, there is the rickshaw economy, as much an animating force in urban areas as the search for usable soil is in villages. Dhaka alone, a city of more than 10 million people, has several hundred

thousand bicycle rickshaws. A rickshaw driver generally pays a rickshaw *mustan* (a mafia-style gang, often associated with a political party) the equivalent of $1.35 a day to rent the rickshaw. He collects 30 cents from an average passenger and ends up making around a dollar a day in profit. His wife may earn a similar amount breaking bricks into road material, while their children sift through garbage.

The slums of the world teem with invisible private schools of just a few teachers selling their services directly for pennies. As Clive Crook reported in the *Atlantic*, "The parents of full-fee students, desperately poor themselves, willingly subsidized those in direst need. On the whole, dime-a-day for-profit schools are doing a better job of teaching the poorest children than the far more expensive state schools."

This book was finished before the full effects of the world financial crisis of 2009 could be studied. Did the informal economy in the world's slums offer a refuge from it, or did people there suffer more than anybody? (A 2009 report in the *Wall Street Journal* said that in the developing world, "people are landing in the informal sector, which has become a critical safety net as the economic crisis spreads.") Were some people driven further into the crime economy? For that matter, how did global crime do? Did the rate of urbanization slow down or speed up? Here's my prediction from March 2009: significant growth in the informal economy from more people retreating into it; significant growth in the crime economy, because it's always ready to exploit chaos; no change in the rate of urbanization. How wrong was I?

• Infrastructure may be hidden in the formal world, but not in squatter cities. Wires are strung everywhere, carrying pirated electricity and cable TV; plastic pipes snake in all directions, illegally distributing water. It is do-it-yourself infrastructure. AES, a global energy company serving the developing world, invited me to a conference in Buenos Aires in 2006. The company was trying to convert electricity pirates into paying customers in Latin American slums, beginning with Caracas in Venezuela. They said it took months with the slum dwellers to build trust, because "they've been lied to by everyone"; but once the company had their trust, it was golden. Bootleg electricity, they learned, was annoyingly unreliable for the users, and it was "dirty": its wave variability would routinely fry TVs and refrigerators. Also it was dangerous; four people a month were fatally electrocuted messing around with the neighborhood's amateur power lines.

"There's plenty of money in the slum," AES people told me, but for any particular household, the income is too irregular to manage a monthly electric bill. So AES reinvented the token electrical meter, similar to the British system used after World War II. When you have some money, you buy some of the coinlike tokens and feed them, as needed, into your meter for a set amount of clean, reliable electricity. To promote solidarity with the Caracas slum community, AES hired some of the more skilled pirates to help install the system. AES was trustworthy, and its new customers were trustworthy, but President Hugo Chávez turned out not to be. After initially supporting the effort, he nationalized the system in 2007 and threw out AES. (The company took its lessons learned and its Caracas team to install a similar system in São Paulo.)

AES, among many other corporations, was inspired by C. K. Prahalad's *The Fortune at the Bottom of the Pyramid: Eradicating Poverty Through Profits* (2005). The book spells out how companies can reach the world's 4 billion poor and deliver goods and services at the interface between the formal and informal economies. Prahalad writes that the poor shop after seven P.M.; they buy in tiny quantities; they welcome relief from the premium prices they often have to pay to slum (sometimes criminal) monopolies; and they are comfortable leapfrogging to new technologies.

Corporations are right to pay attention to what is now referred to as the BOP—bottom of the pyramid. A large portion of humanity on the loose, trying new things in new cities, is a lot of potential customers, collaborators, and competitors, and while the income of the poor is currently small, it is growing fast. And the aggregate numbers are formidable. A 2007 book, *The Next Four Billion*, declares:

> The 4 billion people at the base of the economic pyramid—all
> those with incomes below $3,000 [per year] in local purchas-

ing power—live in relative poverty. Their incomes in current U.S. dollars are less than $3.35 a day in Brazil, $2.11 in China, $1.89 in Ghana, and $1.56 in India. Yet together they have substantial purchasing power: the BOP constitutes a $5 trillion global consumer market.

And an enormous labor market. Researchers for *The Challenge of Slums* were surprised to discover that global corporations, thanks to global oversight by home governments and NGOs, often provide the best-paying jobs and best working conditions in developing-world cities, raising the standard for everybody.

• Everyone in an informal economy lives in the gray area between legitimacy and criminality; over time they can go either way. Smart governments, NGOs, and corporations make it easy for people to move gradually toward legitimacy, because when they are more easily drawn into the criminal economy, it can be a disaster. A sense of the immense scale and damage of the criminal market can be found in the book *Illicit* (2005), by Moisés Naím. The total global traffic in drugs, arms, illegal workers (including sex workers), protected animals and plants, looted artifacts, stolen intellectual property, illegal money flows, etc., adds up to between $1 trillion and $3 trillion a year. The informal economy in the world's slums provides superb cover and a huge workforce for the growth of crime.

Many slums are governed by local organized-crime groups—sometimes responsibly, sometimes destructively, usually a combination of the two. One of the most pathological situations in the world developed in Brazil, where drug lords long ran the *favelas* in a relatively benign way except when they battled with police or each other. Banks in the *favelas* were never robbed, for example. But in the past decade, a new underground force of national scale has emerged: the Primeiro Comando da Capital, or PCC. It is a large, highly disciplined group run from inside Brazil's prisons via cellphones, capable of massive swarming attack. In May 2006 and again in July, the PCC paralyzed the entire city of São Paulo with a series of coordinated violent attacks. Why? Just to prove they could, apparently.

William Langewiesche, writing in *Vanity Fair*, saw the PCC as part of a much larger phenomenon, which he called the "feral zone":

> That zone is a wilderness inhabited already by large populations worldwide, but officially denied and rarely described. It is not a throwback to the Dark Ages, but an evolution toward something new—a companion to globalization, and an element in a fundamental reordering that may gradually render national boundaries obsolete. It is most obvious in the narco-lands of Colombia and Mexico, in the fractured swaths of Africa, in parts of Pakistan and Afghanistan, in much of Iraq. But it also exists beneath the surface in places where governments are believed to govern and countries still seem to be strong.

The PCC may control half the slums in Brazil, but yet another force has recently emerged—militias organized as vigilante groups within some *favelas*. Armed combat goes on among drug lords, the PCC, the police, and now the militias. In her thirty-year study of Rio de Janeiro, Janice Perlman observed: "In 1969, people were afraid of their homes and communities being removed by the government. Today they are afraid of dying in the crossfire between drug dealers and police or between rival gangs. . . . More than one in four people (27 percent) said that some member of their family had died in a homicide."

People who make their life in a feral zone can take no account of environmental issues or climate change. If the feral zone is still growing when climate change begins to disrupt societies, a feral response could multiply the problems and lead to social-chaos wars as well as resource wars. We would be back in the world of "constant battles," at multiple levels.

● It doesn't have to go that way. "San Francisco used to be a shantytown," Rob Neuwirth pointed out to a San Francisco audience. All the world's cities were shantytowns in their early days. The process by which they became proper cities is being recapitulated now in the world's squatter cities, only much faster this time and on a far larger scale. All that the keepers of the formal economy have to do, Neuwirth says, is meet the squatters halfway—help them secure their tenure and give them time to gradually join the formal world, which will no doubt be reshaped by their joining it. Program by program, nation by nation, the world is learning how to engage the boundless resourcefulness of urban squatters.

One idea put forward early was Hernando de Soto's. In his seminal 1989 book, *The Other Path*, he was the first to honor the way the informal economy works, based on research in the squatter communities of Lima. His theory was that squatters could break out of poverty if only they could get bankable title to their shacks. The idea has been tried, but it mostly doesn't work in urban slums: The practice doesn't help with credit, it encourages building owners to become absentee landlords, and it attracts rich raiders. Neuwirth wrote in *Shadow Cities:*

> It doesn't matter whether you give people title deeds or secure tenure, people simply need to know they won't be evicted. When they know they are secure, they build. They establish a market. They buy and sell. They rent. They create. They develop. Actual control, not legal control, is the key. Give squatters security and they will develop the cities of tomorrow.

The lesson I draw is that any practice that leads to treating houses mainly as property tends to destroy community, and any practice that treats them mainly as homes preserves community.

The transition from slum to ordinary urban neighborhood goes best if it is gradual and carefully finessed to local conditions by all the players— dwellers, government, NGOs, and companies. According to a 2008 article in Toronto's *Globe and Mail*:

> "The best plans generally let the slum dwellers themselves make the main decisions in planning their future. You should provide clean water, toilets, electricity, garbage collection and disposal, and maybe let people build their own houses if they can, using materials that you can provide," says Aprodi- cio Laquian, the Filipino-Canadian planner who practically

invented the idea of slum-dweller-designed urban rehabilita-
tion in the 1960s. . . . These sorts of schemes, known as "slum
upgrading" or "sites and services," have been at the heart of
the most successful urban-renewal projects of the past 40
years.

As the 2003 UN report points out, "When more than half of the urban
population lives in them, the slums become the dominant city." A city
planet has every reason to learn to understand slums, to respect the people
there, and to help clear the way for them to become full citizens. That in
turn helps the world, for practical as well as ethical reasons. There's more
to be said about what the world gains from its urban majority, and how.

Footnotes for this chapter may be found at **www.sbnotes.com**.

3

Urban Promise

The city is all right. To live in one
Is to be civilized, stay up and read
Or sing and dance all night and see sunrise
By waiting up instead of getting up.

—Robert Frost, *A Masque of Mercy*

C ities accelerate innovation; they cure overpopulation; and while they are becoming the Greenest thing that humanity does for the planet, they have a long way to go.

● Motivation plus ingenuity, when manifest in a hell of a lot of newly urban people tired of being poor, drives innovation of world-changing originality and scale. The textbook case is what's been happening with cellphones lately.

When the BBC sent a reporter and film crew to Kenya in 2007, they found a cellphone-summoned flash mob of squatters surrounding bull-dozers that were about to force evictions. They found a farmer using what he calls his new middleman—his cellphone—to compare tomato prices in several remote towns. They found cattle ranchers checking with their far-flung herders by cell. A young woman in Nairobi said her cell was now her office: "It allows people to be able to own themselves," she said. "It's as big as fire, the wheel, and the railway." Cellphone text messaging is spreading literacy. Cellphone directness works around government graft. Economic development and social change follow the cell towers wherever they go.

The poor of the developing world have turned cellphones into currency. Technology researcher Alex Pentland at MIT reports:

> In some parts of Africa and south Asia, banking is done by moving around the money in cellphone accounts and people pay for vegetables and taxi rides by SMS [Short Message Service—texting]. Because remanufactured cellphones cost $10 in the developing world and incoming messages are free, every stratum of society is connected. Day laborers, for instance, no longer hang around a street corner waiting to be picked for work. Instead, job offers arrive by SMS from a computerized clearing house. The International Telecommunications Union estimates that in the poorest countries each additional cellphone installed adds $3,000 to the GDP, primarily due to the increased efficiency of business processes.

The great monetizer is prepaid SIM (Subscriber Identity Module) cards—a business worth $3 billion across Africa in 2007, run by local entrepreneurs. SIM cards are sold at vegetable stands and the like, along with "scratch cards" that reload the SIM with additional minutes. Some people carry just a card and borrow a phone when needed. Safaricom, in Kenya, has a service called M-Pesa that lets the cell work as an ATM; to send someone money, you text-message the appropriate code to them, and they get cash from a local M-Pesa agent. Cellphone minutes are traded by phone as a cash substitute. Credit card payments are made by cellphone. Remittances from relatives overseas come by cellphone. (Amounting to about $350 billion a year these days, remittances are expected to reach $1 trillion soon; in some developing countries, the remittance total is already higher than foreign aid and foreign investment combined.)

A South African company called Wizzit offers bankless banking via cellphone, with services I wish we had in California. You can open an account in two minutes anytime, anywhere. You can transfer money from your Wizzit account to virtually any other bank account, pay your bills with it, and get cash at any ATM worldwide. Its service centers are multilingual. Children can have accounts. Services similar to those of M-Pesa and Wizzit are offered by GCash in the Philippines.

There are cellphones with software for Muslims: Five times a day the cellphone calls the user to prayer and then shuts itself off for twenty minutes. An NGO in the Republic of the Congo provides smart phones "to local teachers, elders and business leaders so that they can report incidents of children being drafted as soldiers." Some phones include games that teach languages. Nokia is producing cellphones that have seven address books, for multiple users.

Cellphones are demonstrating how continent-scale "incremental infrastructure" can be built, writes Harvard-based blogger Ethan Zuckerman. All an entrepreneur needs is one cell tower, which can be financed by selling handsets to local customers. As revenue comes in, another tower is built. Web connectivity can be added, one satellite dish at a time. Before long, the locals can make money doing data entry for some American company or by providing translation service. The proliferating cell towers and cellphone battery chargers need electricity, which is currently supplied with expensive inefficiency by diesel-fueled generators. Zuckerman expects incremental electricity microgrids to follow the cell towers.

● For his fine book on the origins of the cellphone revolution, *You Can Hear Me Now* (2007), Nicholas Sullivan spoke with Iqbal Quadir, founder of MIT's Legatum Center for Development and Entrepreneurship. Quadir, who grew up in a Bangladesh village, was working in a New York finance office when his computer network went down. "I realized that connectivity *is* productivity, whether it's in a modern office or an underdeveloped village," he recalls. So in 1994 Quadir set about building a cellphone network for Bangladesh. Allying with the Grameen Bank—the microfinance organization that earned Muhammad Yunus the Nobel Prize in 2006—he seeded the country's towns and villages with "phone ladies" who rented out time on their cellphones to neighbors. By 2008 Grameenphone had over $1 billion in revenues, and Iqbal's brother, Kamal Quadir, had created a cellphone-based virtual marketplace for Bangladesh. Called CellBazaar, it enables massive direct commerce for goods and services costing as little as two cents. "Grameen Bank has an impact on the poor, Grameenphone on the entire economy," said Muhammad Yunus. Sullivan's book chronicles similar revolutions in Africa, in India, and in the Philippines.

The boundless resourcefulness of the poor is evident in slides shown at conferences by Jan Chipchase, Nokia's field ethnologist of tech improvisation in the developing world. He has photos from Indian slums of whole streets of cellphone repair shops, where local talent has reverse-engineered every handset in the world and produced illustrated repair manuals in the local language. One of his slides shows a shabby doorway in a Uganda slum. Over the turquoise-painted door is a hand-lettered "077399721." It's not a street address, it's a cellphone number. The world's slums are the first urban environments to shape themselves around cellphones.

• The bottom of the pyramid is no longer just an overlooked market; it is a creative engine for national economies. "From essentially zero," writes Joel Garreau at the *Washington Post*, "we've passed a watershed of more than 3.3 billion active cellphones on a planet of some 6.6 billion humans in about 26 years. This is the fastest global diffusion of any technology in human history. . . . Cellphones are the first telecommunications technology in history to have more users in the developing world than in the West." American travelers are often shocked to find that cellphone connectivity is better in developing countries than in the United States.

Technology historian Kevin Kelly draws an interesting conclusion from this:

> A decade ago many folks who like to worry about the advance of technology were worried about the "digital divide." This phrase signified the unfair gap between those who had computers and the internet and those who did not. The question was usually framed in these words: "What are you going to do about the digital divide?"
>
> At the time my standard reply was, "Nothing. This is a case of the haves and have-laters. The haves (that's us) are going to overpay for crummy early technology that barely works in

order to make it cheaper and better for the have-laters, who will get it for dirt cheap pretty soon." I then went on to say what I still believe: "The have-laters are going to adopt this technology so fast and so widely that very soon all 6 billion people on earth are going to be wired up, and the real thing we should be worried about, if you want to worry, is: What will happen when we are all connected?"

✦

Something similar has happened with fears about overpopulation. City dwellers have few children—the billion squatters the same as everybody else. Thanks to that by-product of urban growth, the core environmentalist panic about overpopulation is quietly being undermined, but the news hasn't gotten around. The impact is so profound that the history of worry about human overpopulation is worth reviewing in light of its new punch line.

The oldest story in the world, the *Epic of Gilgamesh*, was set down in cuneiform four thousand years ago. It celebrated what has been called the first city in human history—the Sumerian capital, Uruk. Not coincidentally, Uruk was the first great center of writing. In the epic, a terrible flood wipes out every human and every creature except for those saved in an ark by the heroic Uta-napishtim. The Babylonian flood had a different cause than Noah's flood in the Bible, which was written more than a thousand years later. David Damrosch writes in *The Buried Book* (2007):

> The underlying problem was not human iniquity, as in the Bible, but the fact that the human race was multiplying uncontrollably, and people had begun making too much noise. Disturbed in their sleep, the gods appeal to their leader, Enlil, who sends the Flood in response. Enlil's action is violent, but it has a certain ecological logic: the noisiness of the human race is an outgrowth of overpopulation, a serious issue in ancient Mesopotamia, whose large populations often put the region's resources under stress.

It all reads like an early chapter in Steven LeBlanc's chronicle of "constant battles" brought about by fecund humanity perpetually colliding with carrying capacity. Thomas Malthus told the same story in *An Essay on the Principle of Population* (1798); so did my teacher Paul Ehrlich, whose book *The Population Bomb* (1968) put overpopulation at the top of the Green agenda. His book begins: "The battle to feed all of humanity is over. In the 1970s and 1980s hundreds of millions of people will starve to death in spite of any crash programs embarked upon now." It concludes with Ehrlich recommending "compulsory birth regulation," including government-provided sterilants in water and staple foods.

● What environmentalists did with the population issue I can illustrate with a personal account. In 1969 Ehrlich's book inspired me to organize a public piece of reality theater, a hunger show we called Liferaft Earth. Modeled on the civil rights sit-ins of the time, it was to be a "starve-in." Well-fed Americans would starve for a week voluntarily, to draw attention to all those in the world starving involuntarily and the millions more who would soon do so because of overpopulation. The "liferaft" was a parking lot surrounded by an inflated tube fence, with a door opening outward with the words "Are you ready to die?" on it. The conceit of the exercise was that all those still fasting inside the liferaft were alive, but they could abandon the ordeal at any time, exit through the door, and "die."

Sixty or so volunteers, mostly from the traveling performance commune called the Hog Farm, took part. The media showed up, hoping for an amusing or instructive story; no one knew how it would play out. The starvers learned that being stuck somewhere with water but no food was at first deeply boring and then deeply enervating. You sank into yourself, guarding every scrap of dwindling energy. To keep our spirits up and our perspective focused, at suppertime the Hog Farm's leader, Wavy Gravy, would recite over a loudspeaker in an unctuous-waiter voice the courses of the fancy dinner we were not eating.

A few sailed through it. Athletic yogis grew bright-eyed "eating the Void," as they said. Most of us suffered. Many "died." At the end of the endless week, the survivors gathered in a big circle for a discussion full of fiercely emotional crosscurrents. A chant emerged. At first it was a reverent "Omm," but it grew into a groan and then became a vast cry of our pain and global pain.

• Paul Ehrlich himself was involved in the next chapter. In 1972 the United Nations convened the first global forum on environmental issues in Stockholm. By then the bestselling *Last Whole Earth Catalog* had brought a million dollars to Whole Earth's nonprofit parent, Point Foundation. Point spent one of its first and biggest grants to send a contingent of malcontents to Stockholm. The group included poets (Gary Snyder, Michael McClure), Indians (especially Hopi elders protesting coal mining at Black Mesa), San Francisco Greensters such as the pioneer whale-saver Joan McIntyre and overpopulation-spokeswoman Stephanie Mills, and the ever-ready Hog Farm, with their buses. One of the buses paraded through Stockholm dressed as a whale. When the tape recorder to play humpback whale songs broke down, the accompanying marchers sang in whale.

A feud about how to deal with overpopulation surfaced in Stockholm, between Ehrlich and his nemesis, Barry Commoner, whose popular book, *The Closing Circle* (1971), directly criticized Ehrlich's population-bomb thesis. Both were on panels in Stockholm, with Commoner slyly planting invidious questions aimed at Ehrlich among various Third World participants in the conference, and Ehrlich yelling back. Commoner's argument was that population policies weren't needed, because what was called "the demographic transition" would take care of everything—all you had to do was help poor people get less poor, and they would have fewer children. Ehrlich insisted that the situation was way too serious for that approach, and it wouldn't work anyway: You needed harsh government programs to drive down the birthrate. The alternative was overwhelming famines and massive damage to the environment.

I was for Ehrlich and against the ecosocialist Commoner. But Ehrlich's predicted famines never came, thanks largely to the green revolution in agriculture, nor did the need for harsh government programs. Instead,

Commoner's thesis of demographic transition turned out to be mostly right, though in a way unanticipated by him or anybody else.

• For decades, environmentalists organized their fears around the prospect of perpetually soaring human population, and early estimates from UN researchers reflected those fears. A population *optimist*, Wallace Kaufman, wrote in 1994: "Fortunately, population growth is likely to level off between 12 and 15 billion midway through the next century." In the apocalyptic, highly influential 1972 book, *Limits to Growth*, the authors foresaw exponentially increasing population reaching a condition of "overshoot" of Earth's carrying capacity, followed by a population crash. That expectation still prevails among many Greens.

The theory's Malthusian premise has been proven wrong since 1963, when the rate of population growth reached a frightening 2 percent a year but then began dropping. The 1963 inflection point showed that the imagined soaring J-curve of human increase was instead a normal S-curve. The growth rate was leveling off. No one thought the growth rate might go negative and the population start shrinking in this century without an overshoot and crash, but that is what is happening. (Of course, if climate change becomes catastrophic, the *Limits to Growth* overshoot curve will be duplicated, not by population smashing through carrying capacity but by a collapsing carrying capacity smashing down population.)

By the late 1990s, I knew something strange was going on with population and birthrates, but I didn't know exactly what until I read "The Global Baby Bust," a 2004 article by Phillip Longman in *Foreign Affairs*. "Some 59 countries, comprising roughly 44 percent of the world's total population, are currently not producing enough children to avoid population decline," Longman wrote. "The phenomenon continues to spread. By 2045, according to the latest UN projections, the world's fertility rate as a whole will have fallen below replacement levels." In the article and in his 2004 book, *The Empty Cradle*, Longman explained the cause: "As more and more of the world's population moves to urban areas in which children offer little or no economic reward to their parents, and as women acquire economic opportunities and reproductive control, the social and financial costs of childbearing continue to rise."

Development expert Paul Polak points out that the need for extra children among the rural poor is thoroughly rational:

> A one-acre-farm family in Bangladesh needs three sons to get ahead—one to help with the farm, one to get a good enough education to land a government job capable of supporting the family from small bribes, and one to get a local job that pays enough to keep his brother, the one aiming for a government job, in school. But to end up with three sons means having eight babies, two of which are likely to die before the age of five, leaving three boys and three girls.

What happens in the cities is described by George Martine in the 2007 UN report *Unleashing the Potential of Urban Growth*: "In urban areas, new social aspirations, the empowerment of women, changes in gender relations, the improvement of social conditions, higher-quality reproductive health services and better access to them, all favour rapid fertility reduction." In the village, every additional child is an asset, but in the slum, every additional child is a liability, so the newly liberated women in town focus on education and opportunity—on fewer, higher-quality children.

That's how urbanization defused the population bomb.

• The magic number is 2.1. If every woman in the world has 2.1 children, on average, then the growth rate is exactly zero; population neither grows nor shrinks. (The number is 2.1 instead of 2 because some children die before they reach the age of reproduction.) Just as the population exploded upward exponentially when the birthrate was above 2.1, it accelerates downward exponentially when it's below 2.1. Compound interest cuts both ways. Fewer children make fewer children, who make fewer children.

Demographics, they say, is destiny. Everybody gets born, moves around, and dies. The statistics of those three events define our world.

Demographically, the next fifty years may be the most wrenching in human history. Massive numbers of people are making massive changes. Having just experienced the first doubling of world population within a single lifetime (3.3 billion in 1962, 6.6 billion in 2007), we are discovering that it was the last doubling. Birthrates worldwide are dropping not only much faster than expected, but much further. It used to be assumed that birthrates would get down to the replacement rate of 2.1 children per woman and level off, but in most places the birthrates continue to dive right on past that point with no bottom in sight. Meanwhile the "population momentum" of the already born and their kids will carry world population to a peak around midcentury and then head downward.

How high a peak? The UN's "median" projection in 2008 was that population would reach a little over 9 billion, but that figure is based on the assumption that birthrates in developed countries will start rising again for some reason. Because that seems unlikely, I think a more probable peak will be 8 billion, followed by a descent so rapid that many will consider it a crisis.

For every woman you know (or are) who has no children, some other woman has to have 4.2 just to keep the population even, and they don't. Geneticist William Haseltine puts it harshly: "There's a very odd phenomenon which seems to be a cultural invariant: once women gain economic independence, they do not reproduce our species." In most cases, just the *prospect* of economic independence does the trick, and that's what moving to cities provides.

Because urbanization is currently taking place most rapidly in developing countries, the drop in birthrate is most rapid there, which means those nations are aging the most rapidly, though the effects won't be felt for a while. In Mexico the birthrate dropped from 6.5 in the 1970s to about 2 in 2008, and it is still falling. By midcentury, Phillip Longman predicts, "Mexico will be a less youthful country than the United States." Fertility rates are falling so fast in the Mideast and India, he says, that those regions are aging at three times the rate of the United States. China—now at a 1.73 birthrate—is headed toward "demographic meltdown" with its urbanization and one-child policy, because "by 2020 its labor supply will be shrinking and its median age will be older than that of the United States. By midcentury,

China could easily be losing 20–30 percent of its population per generation." Even now, Chinese families complain about the "4:2:1 phenomenon." Each couple finds itself solely responsible for the care of four parents and one child. The prospect, Longman says, is that developing nations may get old before they get rich, and that is its own poverty trap.

● The countries that got rich before they got old are in increasing trouble themselves. The average birthrate in long-urbanized developed nations is down to 1.56 children per woman; in some places it's below 1.2. Those are extinction numbers. It is already in the cards that Russia, Japan, South Korea, Italy, Spain, and Germany will have fewer people in 2050 than they do now. And by then the majority will be old: past childbearing, past being economically productive, with few or no children to care for them, stuck in a national economy that is expiring for lack of young workers.

Longman points out that in Italy, with its birthrate of 1.2, by midcentury, "almost three-fifths of the nation's children will have no siblings, cousins, aunts, or uncles—only parents, grandparents, and perhaps great-grandparents." Instead of having children, Italians are buying pets. So is everyone in the developed world.

How did Japan get itself into a seemingly permanent recession after the dazzling prosperity of the 1980s? Longman told a San Francisco audience:

> Japan boomed through the end of the 1980s, so long as declining fertility was still increasing the relative size of its working-age population. . . . Japan's long recession began just as continuously falling fertility rates at last caused its working-age population to begin shrinking in relative size.

Because Japan welcomes no immigrants, it is facing the world's worst elder-care crisis. At Global Business Network, we predict that Japan's standard solution to labor problems will be applied. Highly sophisticated, lovable robots and robotic environments will take care of Japan's elderly, and then the technology will spread to the rest of the developed world.

The nation with the most alarming outlook is Russia. The birthrate is 1.14. The average Russian woman has seven abortions. One of the demographers at Global Business Network chiseled the nation's epitaph in a 2008 report:

> There is no historical precedent for a national population which is simultaneously aging and shrinking, as is occurring today in Russia. Ecological examples suggest that this demographic trend produces a literal death spiral which is impossible to arrest. Russia entered the "aging and shrinking" situation in the 1990s as a product of a collapsed economy, the rise of HIV transmission from intravenous drug use, major environmental toxification, and extremely poor personal health habits. In Russia, adult male mortality in every age group from 15 to 64 has increased 40 percent-plus since the end of Communism.

● Environmentalists have every reason to rejoice at the defusing of the population bomb, because the aggregate human impact on natural systems, including the atmosphere, could be going down fast. Brazil's birthrate of 1.3 children per woman, for example, may be the best protector of the Amazon rain forest. A population with that birthrate halves in forty-five years, and then halves again in the next forty-five years. In one formulation, that means more resources per person—good news both for the people and the resources. But from another angle, it could mean perpetual economic crisis, which would be terrible news for the environment. In an economic crisis, there is neither money nor attention for responsible stewardship. There is no long-term thinking or action. Wars become more likely, and wars are deadly for the environment.

An enlightened environmental program on population should now focus, I suggest, on softening the impact of the depopulation implosion. Greens can take a bow for dramatizing the importance of population early and for promoting the education, birth-control techniques, and prosperity that helped reduce birthrates worldwide. Hats off to Paul Ehrlich for one of the great self-defeating prophecies in history. And now it's time for follow-through, for a nuanced shift. The most effective environmental population program in this century is gently pronatal.

No one knows how that would work in most countries, however. In Europe, birthrates have been declining steadily for four decades. How do you reverse that? Pope Benedict complained in 2006, "Children, our future, are perceived as a threat to the present." Governments have tried everything to encourage more children. The Australian government has a three-child policy—"One for mum, one for dad, and one for the country"—and its birthrate remains at 1.8. Singapore's advice flipped from "Stop at two" to "Three children or more if you can afford it," and the birthrate there remains one of the lowest in the world at 1.04. All the developed nations of the world are running out of babies and headed for economic decline—with two instructive exceptions.

The United States and France have the highest birthrates in the developed world, just below replacement level. America does it with immigrants and churchgoers. First-generation immigrants have strong family values and large families. (That will dry up as the source countries age quickly and cease to export young people.) Devoutly religious couples, who abound in the United States, pay more attention to God's instruction to be fruitful and multiply than to the heavy economic and opportunity costs of large families.

France does it with socialism. Every mother gets maternity leave at nearly full pay—twenty weeks for the first two children, forty weeks for the third child—and her employer has to keep her job open. Fathers get paid paternity leave. Child care is free. Nursery school is free. Large families get tax credits and free public transport. Parents of a third child get a government bonus of $1,500 a month for a year. France's birthrate rose from 1.92 in 2005 to 1.98 in 2007.

• For the next three decades, the world will be demographically split: in the global north, old cities full of old people; in the global south, new cities full of young people. In the north: slow economic growth, stagnation, or decline; in the south: economic opportunity. Before 2009, everyone marveled at China's 10 percent annual growth; but India had 8 percent, and

Africa—no longer the basket case people think it is—had 7 percent, led by South Africa. In 2007 the U.S. economy grew by 2.2 percent, France's grew by 1.8 percent, and Japan's grew by 1.9 percent. It will be interesting to see who recovers first and fastest following the economic disruptions of 2009.

The nations of the global south are coming into their demographic dividend. The new millions of young adults live in cities, voting with their ovaries for fewer children and more opportunity. They are mostly of working age, burdened by relatively few infants and elderly dependents. "Poor countries with low and falling fertility rates are growing wealthier faster than rich modern countries," wrote Ben Wattenberg in his 2004 book on depopulation, *Fewer.* "We live in a youthful world," says a 2006 UN report. "Almost half of the global population is under the age of 24. . . . Fully 85 percent of the world's working-age youth, those between the ages of 15 and 24, live in the developing world."

While the north is slowing down, the south is speeding up. Innovation comes from the young, because, as tech analyst Clay Shirky points out, "they have an advantage in that they don't have to spend a lot of time *un*learning things that are no longer worth knowing." However, the young are also the major perpetrators and victims of violent crime, and they fill the ranks of armies, militias, and insurgencies.

How will demographic change and climate change intersect? A decreasing population could be seen as a form of adaptation to climate change. As carrying capacity drops, human population drops even faster, as if in anticipation. Rather than killing each other, we just make fewer of each other, substituting a lower birthrate for a higher death rate.

A lot will depend on how the global north thinks about the south and how the south thinks about itself. The populations in developing countries are the most vulnerable to climate change and will be the first to suffer from it—from sea level rise, from extreme weather events, and especially from drought. Many in the north will be tempted to dismiss climate disasters in the south as irrelevant rather than as harbingers of what will be coming soon enough in the north. Likewise, some in the south will want to play the victim, assigning all blame to the wasteful, greedy, colonialist, carbon-spewing north.

A preferable outcome is that the young, urbanized south cranks up its rate of innovation and figures out its own ways to manage climate change, ideally with close engagement from the north. A wide array of best practices for climate adaptation could emerge in the south and gain global application.

◆

Squatters already have inspired some Green practices. There should be many more to come. I'll give one example from close to home. I had been studying and admiring the world's squatter communities for two years before I noticed that I live in one.

Liberty ships, the torpedo fodder of World War II, were built on the Sausalito, California, waterfront at a peak rate of one a day in 1944. When the war was over, the former shipyard became a semioutlaw area, and riff-raff moved in—floated in. Steadily, through the 1950s and 1960s, the tide-lands filled with floating shacks, derelict boats, and habitable sculpture, occupied by artists, maritime artisans, and other people who had more nerve than money.

A benevolent landlord went along with the game—collecting rent casually, protecting the community from outraged government authorities. Electricity was stolen from shore via extension cords, water via garden hoses. People crapped in the bay, and it smelled rank at a minus tide. There was some drug trade and the occasional murder. There was some freelance prostitution ("hitchhookers"). There was also a fashion shop, and a fashion show (of recycled clothes), several rock bands, and a theater group—Antenna—that earned a world reputation. I've lived and worked in the scene since 1973.

From time to time, the Coast Guard or the county sheriff would try to tow away the houseboats. The Sausalito City Council sent a demolition crew on a dawn attack. A state agency created by environmentalists, the Bay Conservation and Development Commission, declared that house-boats are illegal "bay fill" and have been trying to expunge us for thirty-five years. The floating community hired lawyers, survived, grew, and gradually gentrified.

The four hundred or so Sausalito houseboats are mostly legal now, pay rent (berthage), and are hooked up to city infrastructure so the mud smells like mud again. Tourist groups stroll the docks to admire our colorful lifestyle. Chances are you've come across Sausalito waterfront creativity in the writings of Annie Lamott, Alan Watts, Paul Hawken, or Green architect Sim Van Der Ryn; in the cartoons of Shel Silverstein or Phil Frank; in Otis Redding's "Sittin' on the Dock of the Bay"; in the Antenna Theater–produced Audio Tours that guide you around the world's museums and historic sites; in the biological paintings of Isabella Kirkland; and in any town or city reshaped by what is called New Urbanism. That last item is my example.

● In 1983, architect Peter Calthorpe gave up on San Francisco, where he had tried and failed to organize neighborhood communities, and moved to a houseboat on the end of South Forty Dock, where I live. He found he was in a place that had the densest housing in California, where no one locked their doors—where most of the doors didn't even have locks. Without trying, it was an intense, proud community. When Calthorpe looked for some element of design magic that made it work, he decided it was the dock itself, and the density. Everyone in the forty-nine houseboats on the dock passed each other on foot daily, trundling to and from the parking lot on shore. Everyone knew each other's faces and voices and cats. It was a community, Calthorpe decided, because it was walkable.

Building on that insight, Calthorpe became one of the founders of New Urbanism, along with Andrés Duany, Elizabeth Plater-Zyberk, and others. In 1985 he introduced the concept of walkability in "Cities Redefined," an article in the *Whole Earth Review*. Since then, New Urbanism has become the dominant force in city planning, promoting high density, mixed use, walkability, mass transit, eclectic design, and regionalism. It drew one of its major ideas from a squatter community.

There are a lot more ideas where that one came from. For instance, shopping areas could be more like the lanes in squatter cities, with a dense interplay of retail and services—one-chair barbershops and three-seat bars interspersed with the clothes racks and fruit tables. "Allow the informal sector to take over downtown areas after 6 p.m.," suggests Jaime Lerner, the renowned former mayor of Curitiba, Brazil. "That will inject life into the city." In the thousands of squatter cities in the world, a billion creative people, most of them young, are trying new things unfettered by law or tradition.

Squatter cities are Green. They have maximum density—a million people per square mile in Mumbai—and minimum energy and material use. People get around by foot, bicycle, rickshaw, or the universal shared taxi variously called a *matatu* (Kenya), *dala-dala* (Tanzania), *tro-tro* (Ghana), *jeepney* (Philippines), *tuk-tuk* (Thailand), *tap-tap* (Haiti), *maxi-taxi* (Romania), etc. (Not everything is efficient in the slums, though. In the Brazilian *favelas* where electricity is stolen and therefore free, Jan Chipchase from Nokia found that people leave their lights on all day.)

In most slums recycling is literally a way of life. The Dharavi slum in Mumbai has four thousand recycling units and thirty thousand ragpickers; six thousand tons of rubbish are sorted in the slum every day. In Vietnam and Mozambique, an article in the *Economist* reports, "Waves of gleaners sift the sweepings of Hanoi's streets, just as children pick over the rubbish of Maputo's main tip. Every city in Asia and Latin America has an industry based on gathering up old cardboard boxes." There's a whole book on the subject: *The World's Scavengers* (2007), by Martin Medina. Lagos, Nigeria, widely considered the world's most chaotic city, has an Environment Day on the last Saturday of every month. From seven to ten A.M. nobody drives, and the entire city, including the slums, tidies itself up.

• In his 1985 article that introduced the idea of walkability, Peter Calthorpe made a statement that still jars most people: "The city is the most environmentally benign form of human settlement. Each city-dweller consumes less land, less energy, less water, and produces less pollution than his counterpart in settlements of lower densities." "Green Manhattan" was the inflammatory title of a 2004 *New Yorker* article by David Owen. "By the most significant measures," he wrote,

> New York is the greenest community in the United States, and one of the greenest cities in the world. . . . The key to New

York's relative environmental benignity is its extreme compactness. Manhattan's population density is more than eight hundred times that of the nation as a whole. Placing one and a half
million people on a twenty-three-square-mile island sharply
reduces their opportunities to be wasteful, and forces the
majority to live in some of the most inherently energy-efficient
residential structures in the world: apartment buildings.

But what about the ecological footprint? The idea of measuring environmental impact in notional acres was introduced by Mathis Wackernagel
and William Rees in the 1996 book *Our Ecological Footprint*, as a way to
estimate the resource efficiency of large systems like cities and as a way
to condemn suburban sprawl. The concept has been tremendously useful in shaming cities into better environmental behavior, but comparable
studies have yet to be made of rural populations, whose environmental
impact per person is much higher than city dwellers. Nor has footprint
analysis been applied to urban squatters, who will certainly score as the
Greenest of all.

Urban density allows half of humanity to live on 2.8 percent of the
land. Soon that will be 80 percent of humanity on 3 percent of the land.
Consider just the infrastructure efficiencies. According to a 2004 UN
report, "The concentration of population and enterprises in urban areas
greatly reduces the unit cost of piped water, sewers, drains, roads, electricity, garbage collection, transport, health care, and schools." In the
developed part of the world, cities are Green mainly because they reduce
energy use, but in the developing world, the primary Greenness of cities lies in their ability to draw people in and take the pressure off rural
natural systems.

The Last Forest (2007), a book by Mark London and Brian Kelly
on the realities of the Amazon rain forest, suggests that the nationally
subsidized city of Manaus in northern Brazil "answers the question
'How do you stop deforestation?' Give people decent jobs. If you give
them jobs, they can afford houses; give them houses and their family
has security; give them security and their vision shifts to the future."
One hundred thousand people who would otherwise be deforesting the

jungle around Manaus are prospering in town making such things as cellphones and TVs.

● Environmentalists have yet to seize the enormous opportunity offered by urbanization. Two major campaigns should be mounted—one to protect the newly emptied countryside, the other to Green the hell out of the growing cities. Because cities change constantly anyway, it's not that hard to improve them.

More than any other political entity, cities learn from each other. News of best practices spreads fast. Mayors travel routinely, cruising for ideas in the cities deemed the world's Greenest—Reykjavik, Iceland; Portland, Oregon; Curitiba, Brazil; Malmö, Sweden; Vancouver, Canada; Copenhagen, Denmark; London, England; San Francisco, California; Bahi de Caráquez, Ecuador; Sydney, Australia; Barcelona, Spain; Bogotá, Colombia; Bangkok, Thailand; Kampala, Uganda; and Austin, Texas. Urban ecotourism is a growth industry.

To manage the ecology of cities, we first have to understand it. A 2008 article in *Science* framed the necessary new discipline in meaty language I find delicious:

> Evolving conceptual frameworks for urban ecology view cities as heterogeneous, dynamic landscapes and as complex, adaptive, socioecological systems, in which the delivery of ecosystem services links society and ecosystems at multiple scales. . . .
>
> The changes in chemical environment, exposure to pollutants, simplified geomorphic structure, and altered hydrographs of urban streams combine to create an urban stream "syndrome" of low biotic diversity, high nutrient concentrations, reduced nutrient retention efficiency, and often elevated primary production. . . . Countering the urban stream syndrome may require abandonment of the ideal of a "restored" stream in favor of a designed ecosystem. . . . Reconciliation ecology, where habitats greatly altered for human use are designed,

spatially arranged, and managed to maximize biodiversity while providing economic benefits and ecosystem services, offers great promise that ecologists will be increasingly called upon to help design and manage new cities and reconstruct older ones.

Progress comes from mashup notions like "socioecological systems" and "reconciliation ecology." The new profession of urban ecology could unleash hordes of postdocs on everything from cockroach predation to urban disease vectors and help cities engage natural infrastructure with the same level of sophistication that is brought to built infrastructure. Every Green organization should have an urban strategy, and some should specialize in cities.

One idea that could be transferred from squatter cities is urban farming. Another article in *Science* enthused:

> In a high-tech answer to the "local food" movement, some experts want to transport the whole farm—shoots, roots, and all—to the city. They predict that future cities could grow most of their food inside city limits, in ultraefficient greenhouses. . . . Well-designed greenhouses use as little as 10 percent of the water and 5 percent of the area required by farm fields. . . . A 30-story farm on one city block could feed 50,000 people with vegetables, fruit, eggs, and meat. Upper floors would grow hydroponic crops; lower floors would house chickens and fish that consume plant waste.

Urban roofs offer no end of opportunities for energy saving and "reconciliation ecology." Planting a green roof with its own ecological community is a well-established practice. For food, add an "ultraefficient greenhouse"; for supplemental power, add a few of the current generation of solar collectors.

The most dramatic gains can come from simply making everything white. A white roof saves the building's tenant 20 percent in cooling costs; that's why California now requires all new and retrofitted buildings in the state to have heat-reflective roofs. If you also plant plenty of trees, you significantly reduce the "heat island" effect of a city, which in turn reduces smog. About a quarter of a city's surface is roofs, and a third is pavement, which can be made paler with concrete or with light-colored aggregate in the asphalt. Since a white city reflects sunlight instead of absorbing it, there are climate benefits. According to a 2008 study from the Lawrence Berkeley National Laboratory, "If the 100 largest cities in the world replaced their dark roofs with white shingles and their asphalt-based roads with concrete or other light-colored material, it could offset 44 metric gigatons (billion tons) of greenhouse gases." Call the program "Alabaster Cities," and you can invoke "America the Beautiful" in support. (It's the best line in the song: "O beautiful for patriot dream / That sees beyond the years / Thine alabaster cities gleam / Undimmed by human tears.")

Some environmentalists already are proponents of urban compactness. Sierra Club's magazine reports that in Vancouver, "Mayor Sam Sullivan's EcoDensity program includes zoning changes to allow 'secondary suites,' or in-law apartments; triplexes; and narrow streets with houses that abut property lines." Peter Calthorpe's "walkability" has become a real estate selling point, with walkable neighborhoods able to charge premium prices. A proven way to encourage walking and use of public transit is with a "congestion tax" on cars in the downtown streets. In 2002 London followed the lead of Singapore and Hong Kong and adopted the practice of charging cars £8 a day to drive in the central city. Complaints died away when everybody's travel times in and out of the city went down dramatically. Stockholm, San Francisco, Sydney, and Shanghai are taking up the idea, with more cities to follow.

One abiding problem is that the high cost of living in a city prices most families with children right out of town to the suburbs, and the good schools follow them. Greens could help reverse that trend by pressuring cities to become more like child-friendly Paris, where every neighborhood has excellent schools and parks with playgrounds, puppet theaters, and carousels. Peter Calthorpe tells me he's become an advocate of voucher schools, because that system forces city schools to compete in an open market, and they then improve sufficiently to attract the families back from the suburbs. New forms of subsidized family housing should be explored.

Infrastructure makes cities possible, and it has to be rebuilt every few decades. According to a report in 2007 by the infrastructure consultants Booz Allen Hamilton, "Over the next 25 years, modernizing and expanding the water, electricity, and transportation systems of the cities of the world will require approximately $40 trillion." What would infrastructure totally rethought in Green terms look like? China is currently building 170 new mass transit systems. High-speed rail is finally coming to the United States. With the coming of "smart grids" and microgrids, the distribution of electricity will be reshaped toward greater adaptability as well as efficiency.

As climate change unfolds, cities will be on the frontier of human response. Taking the danger zone as 30 feet above sea level, a Columbia University study reported in *Science* says that two thirds of all cities with a population over 5 million are "especially vulnerable" to rising sea levels and "weather oscillations." The Thames Barrier protecting London from flood tides was raised twenty-seven times between 1986 and 1996, and sixty-six times between 1996 and 2006. Some are forecasting that it will be overwhelmed by 2030.

From the *mitigation* angle, it will be worth refining a "climate footprint" template for cities, grading them on such things as their albedo, their vegetation density, and their greenhouse gas and soot output versus their use of carbon-free energy sources like hydro, nuclear, wind, and solar. As with ecological-footprint studies, there should be a time dimension—is the city improving or getting worse? Then comes *adaptation*. A "climate prospects" template would detail how each city might respond to climate impacts such as sea-level rise, drought, extreme weather, and temperature changes. Is high ground nearby? Is there a move-upstairs option in the buildings? Can the local water supply and agriculture adapt to salt water infusion or, if inland, to drought? If city abandonment becomes necessary, how would it play out?

• In the broad scope of history, growing cities are far from an unmitigated good. They concentrate crime, pollution, and injustice as much as they concentrate business, innovation, education, and entertainment. If they are overall a net good for the people who move there, it is because cities offer more than just job opportunity. They are transformative. In the slums as well as the office towers and leafy suburbs, the progress is from hick to metropolitan to cosmopolitan, and everything the dictionary says that *cosmopolitan* means: multicultural, multiracial, global, worldly-wise, well traveled, experienced, unprovincial, cultivated, cultured, sophisticated, suave, *urbane*.

The takeoff of cities is the dominant economic event of the first half of this century. Among all its other impacts will be infrastructural stresses on energy supply and food supply. People in vast numbers are climbing the energy ladder from smoky firewood and dung cooking fires to diesel-driven generators for charging batteries, then to 24/7 grid electricity. They are also climbing the food ladder—from subsistence farms to cash crops of staples like rice, corn, wheat, and soy to the high protein of meat—and doing so in a global marketplace. Environmentalists who try to talk people out of such aspirations will find the effort works about as well as trying to convince people to stay in their villages did.

Peasant life is over unless catastrophic climate change drives us back to it.

The demographic literature refers often to the "bright lights" phenomenon that draws people to cities. Thanks to military satellite imagery, those lights are now visible to us from space. The night side of Earth, these decades, displays a dazzling lacework of light on the continents, with incandescent nodes at the metropolitan areas and a bright tracery of transportation corridors between them. That web of light is the sign to any visitor that they are approaching not just a living planet, but a civilized planet.

What powers all that light?

Footnotes for this chapter may be found at **www.sbnotes.com**.

4

New Nukes

Coal is the killer. Of all the fossil fuels, coal is the one that could make this planet uninhabitable.

—Fred Pearce, *New Scientist*

With climate change, those who know the most are the most frightened. With nuclear power, those who know the most are the least frightened.

—Variously attributed

For the definitive word on how much to worry about climate change, environmentalists in America have taken to relying on James Hansen, NASA's authoritative and outspoken climatologist. When Hansen declared in 2007 that we must not settle for leveling off carbon dioxide in the atmosphere at 450 parts per million (ppm) but must take the level *down* from the current 387 ppm to 350 ppm or lower, the new environmentalist slogan became "350!"

The respect goes only in one direction in regard to nuclear, however. As President Obama was taking office, Hansen wrote him an open letter suggesting new policy to deal with the climate crisis. "Coal plants are factories of death," he wrote. "Coal is responsible for as much atmospheric carbon dioxide as the other fossil fuels combined." Hansen proposed what America needed: a carbon tax "across all fossil fuels at their source"; the phasing out of all coal-fired plants; and "urgent R&D on 4th-generation nuclear power, with international cooperation." He warned: "The danger is that the minority of vehement anti-nuclear 'environmentalists' could cause development of advanced safe nuclear power to be slowed such that utilities are forced to continue coal-burning in order to keep the lights on. That is a prescription for disaster." He repeated the point at the end of the letter: "One of the greatest dangers the world faces is the possibility that a vocal minority of anti-nuclear activists could prevent phase-out of coal emissions."

Environmentalists have much less to fear in reality from the current nuclear power industry than they think, and much more to gain from new and planned reactor designs than they realize. Hansen is right: Nukes are Green; new nukes even more so. Here's how.

● Nuclear power inspires in most environmentalists one particularly deep aversion. They recoil from the idea of passing on to endless future generations the care of the deadly poison of nuclear waste. That was my view as well until one day in 2002. It was a visit to Yucca Mountain, of all things, that began to change my mind about nuclear power. I'll report the occasion in some detail because it's a look inside a "here-be-dragons" blank area on most people's mental map of the nuclear world, and you'll watch two people reverse their opinion about nuclear and an organization change its mind about itself.

The Yucca Mountain Repository for "spent nuclear fuel and high-level radioactive waste" one hundred miles northwest of Las Vegas, Nevada, has been lodged in America's political throat ever since the project was initiated in 1978. That had nothing to do with why the board of The Long Now Foundation made a site visit in 2002. We just wanted to see what a hole in a Nevada mountain looked like.

Long Now's maypole project (around which everything else dances) is a monumental ten-thousand-year clock, to be installed inside a mountain in eastern Nevada as an icon to, as we say, "help make long-term thinking automatic and common instead of difficult and rare." What kind of spaces work best inside a mountain, we wondered, especially inside a desert mountain? We thought that Yucca might give us some hints, and indeed it did—long, straight, cylindrical tunnels are boring; a twenty-five-foot ceiling is boring; but anything below ten feet is cozy, and anything above thirty-five feet is thrilling.

Our main lesson from Yucca, though, threatened Long Now's very core. Something was pathological about Yucca Mountain, and the sickness was embedded in its long-term thinking, its ten-thousand-year time frame. Among those on the bus were Danny Hillis, designer of the clock, and Peter Schwartz, cofounder of Global Business Network. I wrote in my trip report that at the entrance to the Repository,

> a training video informed us that it was important not to trip on anything, and showed how to use a belt-mounted emergency breathing apparatus. Danny Hillis remarked that it is the device which, in event of a mine fire, OSHA demands to find on your body. Outside the tunnel entrance were not one but two brand new ambulances, contextually shrieking "SAFETY, SAFETY, SAFETY!!"
>
> After a briefing in a pleasant underground "alcove" we rode a noisy train a mile and a half into the mountain—laser straight in a 25-foot diameter tunnel. The overall loop tunnel is 5 miles long. The eventual storage "drifts" have yet to be excavated, except for a few test drifts. We got off the train to visit one of the tests, where a vastly expensive experiment is under way to see how heating and cooling affects the rock and water flow around the drift—four years heating it up as if it had radioactive waste in it, four years cooling it down, modeling the first 1,000 years of waste storage.
>
> That evening we debriefed the Yucca Mountain experience over dinner. We were universally appalled that the government had poured $8 to $16 BILLION ("depending on how you count") into that hole in the ground. Most of the money had gone into gargantuan tests meant to reassure the public that the stored waste would be "safe" for 10,000 years. It was a grotesque expenditure, based on 1950s ideas, a deeply political set of gestures meant to reassure critics who are largely uninterested in science and distrustful no matter what.
>
> Peter Schwartz bet that if the waste goes in the mountain

(there's a 50 percent chance that it will), in 50 to 100 years we will be taking it back out as a valuable energy resource.

I proposed that it was the 10,000-year time frame that made people crazy, which calls into question the whole premise of Long Now. We asked ourselves: If Long Now had been asked to handle the nuclear waste problem, what would we have done? Danny said, "I would have built the same hole in the ground, for only a couple hundred million, and told everyone that we just wanted to put the waste there for a hundred years while we thought about what to do with it eventually."

We realized that Yucca Mountain is a classic example of the folly of long-term planning—the illusion that we know now how to do the right thing for the next ten millennia. What Long Now pushes is almost the opposite: long-term thinking—where you set in motion a framing of events so that a process is made intensely adaptive, preserving and indeed increasing options as time goes by.

The more I thought about the standard environmentalist stance on nuclear waste, which I had espoused for years, the nuttier it seemed to me. The customary rant goes: "You have to guarantee that all the radioactivity in the waste will be totally contained for ten thousand years (no, a hundred thousand years; no, a million years), and if you can't guarantee that, you can't have nuclear power." Why? "Because any amount of radioactivity hurts humans and other life forms. It might get in the ground water."

What humans? The assumption seems to be that future humans will be exactly as we are today, with our present concerns and present technology. How about, say, two hundred years from now? If we and our technology prosper, humanity by then will be unimaginably capable compared to now, with far more interesting things to worry about than some easily detected and treated stray radioactivity somewhere in the landscape. If we crash back to the stone age, odd doses of radioactivity will be the least of our problems. Extrapolate to two thousand years, ten thousand years. The problem doesn't get worse over time, it vanishes over time.

The Yucca trip set in motion another board member's conversion from anti- to pro-nuclear. Peter Schwartz, an energy expert who served for a long time on the board of Amory Lovins's Rocky Mountain Institute, eventually became highly vocal on behalf of a nuclear revival. He and Lovins had friendship-threatening arguments on the subject.

● A year after the Yucca trip, Global Business Network was invited to run a scenario workshop for Canada's Nuclear Waste Management Organization, which was conducting a series of meetings to explore what Canada should do with the waste from its twenty-two CANDU nuclear reactors. One option was to heave it down a deep hole in the ancient, stable bedrock of the Laurentian Shield and forget about it. Another was to leave it where it is now in dry cask storage at the reactor sites. Another was to develop a Yucca-like site for retrievable underground storage. At the workshop, I told my Yucca Mountain story. Also participating in the workshop were several Indians (the tribes are called First Nations in Canada) who proposed taking the "seven generations" approach to future responsibility long credited to the Iroquois League. Using the standard number for a generation—25 years—that would mean a 175-year time frame for thinking about the waste.

After eighty meetings across Canada, the nation's nuclear waste policy emerged. It is based, says a report from the organization, on the principle of "Respect for Future Generations: we should not prejudge the needs and capabilities of the future. Rather than acting in a paternalistic way, we should leave the choice of what to do with the used fuel for them to determine." Accordingly, Canada has an "adaptive phased management" plan, where the spent fuel remains in wet and dry storage at the reactor sites while a "near term" (1 to 175 years) centralized shallow underground facility is built, designed for easy retrieval; that will be followed by a deep geological repository for permanent storage. Future Canadians have options at every step. No mention is made of 10,000 years. The report does note that "during the 175-year period, the overall radioactivity of used fuel drops to one-billionth of the level when it was removed from the reactors." Nuclear waste has the interesting property that it loses toxicity over time, unlike many forms of chemical waste, such as mercury.

Two things about nuclear had changed for me, I gradually realized. Waste disposal no longer looked like a cosmic-level problem, and carbon-free energy from nuclear looked like a major solution in light of growing worries about climate change. My opinion on nuclear had flipped from anti to pro. The question I ask myself now is, What took me so long? I could have looked into the realities of nuclear power many years earlier, if I weren't so lazy.

Gwyneth Cravens, a novelist and former *New Yorker* editor, did what I should have done. In 1980 she was among the activists who shut down the $6 billion Shoreham Nuclear Power Plant in Long Island before it ever opened, which helped frighten the American nuclear industry to a standstill. In the 1990s, she started hearing the other side from a friend in the industry, a nuclear-safety scientist at Sandia National Laboratories in Albuquerque named Rip Anderson. Sensing a story, she traveled with Anderson as her guide through the U.S. nuclear power industry and came up with a masterly account of the journey, *Power to Save the World: The Truth About Nuclear Energy*, published in 2007.

I asked her what really changed her mind about nuclear. "Two things," she said. "Baseload and footprint."

" 'Baseload,' " she explains in the book, "refers to the minimum amount of proven, consistent, around-the-clock, rain-or-shine power that utilities must supply to meet the demands of their millions of customers." Baseload is the foundation of grid power. So far it comes from only three sources: fossil fuels, hydro, and nuclear. Two thirds of the world's electricity is made by burning fossil fuels, mostly coal. The Green, noncarbon one third is split evenly between hydroelectric dams and nuclear reactors at about 16 percent each. (In the United States, 71 percent of our electricity is from coal and gas, 6.5 percent from hydro, about 20 percent from nuclear.)

Cities require grid power, and that means baseload. The world's growing cities and the billions of people climbing the "energy ladder" out of poverty will demand a lot more baseload by midcentury. If climate is the major Green threat, and cities are a major Green boon, then nuclear power looks doubly Green.

Wind and solar, desirable as they are, aren't part of baseload because they are intermittent—productive only when the wind blows or the sun shines. If some form of massive energy storage is devised, then they can participate in baseload; without it, they remain supplemental, usually to gas-fired plants. (Space-based solar, however, could feed directly into baseload, microwaving the juice from orbit down to surface rectennas. Sunlight in space has three times the intensity of the pallid stuff on Earth, and it's always on, so solar panels in space have three times the sun exposure of solar panels on roofs. That adds up to a ninefold advantage. Expensive commute, though. Japan is planning a 1-gigawatt space solar facility nevertheless and a California utility claims it will have a 200-megawatt solar farm in orbit by 2016.)

● As for footprint, Gwyneth Cravens points out that "A nuclear plant producing 1,000 megawatts takes up a third of a square mile. A wind farm would have to cover over 200 square miles to obtain the same result, and a solar array over 50 square miles." That's just the landscape footprint. (By the way, 1,000 megawatts equals 1 gigawatt—a billion watts; I'll use that measure most of the time here.)

More interesting to me is the hazard comparison between coal waste and nuclear waste. Nuclear waste is minuscule in size—one Coke can's worth per person-lifetime of electricity if it was all nuclear, Rip Anderson likes to point out. Coal waste is massive—68 tons of solid stuff and 77 tons of carbon dioxide per person-lifetime of strictly coal electricity. The nuclear waste goes into dry cask storage, where it is kept in a small area, locally controlled and monitored. You always know exactly what it's doing. A 1-gigawatt nuclear plant converts 20 tons of fuel a year into 20 tons of waste, which is so dense it fills just two dry-storage casks, each one a cylinder 18 feet high, 10 feet in diameter.

By contrast, a 1-gigawatt coal plant burns 3 million tons of fuel a year and produces 7 million tons of CO_2, all of which immediately goes into everyone's atmosphere, where no one can control it, and no one knows what it's really up to. That's not counting the fly ash and flue gases from coal—the world's largest source of released radioactivity, full of heavy metals, including lead, arsenic, and most of the neurotoxic mercury that has so suffused the food chain that pregnant women are advised not to eat wild fish and shellfish. The air pollution from coal burning is estimated to cause 30,000 deaths a year from lung disease in the United States, and 350,000 a year in China.

As for comparing full-life-cycle, everything-counted greenhouse gas emissions, a study published in 2000 by the International Atomic Energy Agency shows total lifetime emissions per kilowatt-hour from nuclear about even with those of wind and hydro, about half of solar, a sixth of "clean" coal (if it ever comes), a tenth of natural gas, and one twenty-seventh of coal as it is burned today.

● Baseload. Footprint. Add portfolio—the idea that climate change is so serious a matter, we have to do *everything* simultaneously to head it off as much as we can. The first definitive portfolio statement came from engineer Robert Socolow and ecologist Stephen Pacala in 2004. Their paper in *Science*, "Stabilization Wedges: Solving the Climate Problem for the Next 50 Years with Current Technologies," introduced the idea that a set of "stabilization wedges," made up of already proven technologies and practices, could reduce greenhouse gas emissions to a tolerable level, but only if all the wedges are pursued extremely aggressively at the same time, starting yesterday.

The paper proposed seven wedges to level off emissions: energy efficiency, renewables, clean coal, forests and soils (stop deforestation and agricultural tilling), fuel switch (coal to gas, oil to heat pump, etc.), and nuclear. Tripling the world's current nuclear capacity to 700 gigawatts a year over fifty years would reduce carbon emissions by 1 gigaton a year at the end of the period, for a total of 25 gigatons over the fifty years. (In 2008 the world's carbon emissions were over 7 gigatons a year and rising fast. To convert Socolow's carbon emissions to carbon dioxide emissions—the more common measure—multiply by 3.67; thus the 7 gigatons of carbon equals 25.7 gigatons of carbon dioxide.) There's nothing heroic or unusual about that rate of adding nuclear capacity. The Socolow/Pacala paper noted, "The global pace of nuclear power plant construction from 1975 to 1990 would yield a wedge, if it continued for 50 years."

Al Gore's climate movie, *An Inconvenient Truth*, featured Socolow's diagram with only six wedges, leaving out nuclear.

• Of all the wedges, energy efficiency and conservation come first, last, and always, as far as I'm concerned. You get the most result with the least cost at the greatest speed. As Amory Lovins has been proving eloquently for decades, saving energy doesn't cost money; it makes money, and it can be carried out at every level, from individual behavior to global programs. Those who argued that conservation would do economic harm have been thoroughly refuted by events. Europeans and Californians use half the energy per person of most Americans, and they are doing fine. In the 1970s, the Jerry Brown administration in California set in motion an array of programs that kept energy use dead level for the next three decades while the state's per capita income grew by 80 percent, and while the other states' energy use went up by 50 percent. California's greenhouse gas emissions per capita are less than half of what the other states put out.

How did California do that? A physicist named Arthur Rosenfeld at Lawrence Berkeley National Laboratory was inspired by the OPEC oil embargo of 1973 to shift his career to energy efficiency. Through his advocacy, California pursued "technology forcing" (specific mandates on efficiency in refrigerators and cars); new regulations on heat efficiency in buildings; tax credits for solar panels; and a decoupling of profits from sales in energy companies, so corporations like Pacific Gas & Electric were suddenly motivated to reward energy conservation in their customers and *not* motivated to build new power plants. My gang was in the thick of that. While I served on the governor's personal staff, one Whole Earth friend (Huey Johnson) was secretary of resources, another (astronaut Rusty Schweickart) chaired the Energy Commission, two (Sim Van der Ryn and Peter Calthorpe) were running the State Architect's office, and still another (Wilson Clark) ran the Office of Appropriate Technology .

(Jerry Brown is still at it. Elected attorney general of California in 2006, he immediately focused the office on climate change, with several creative initiatives. One will require that new projects in the state prepare a *climate* impact report as part of their environmental impact report. Another recognizes that the worst polluters of air and atmosphere are the world's totally unregulated ships burning the vilest of fuels, called bunker. Brown set up a coalition of West Coast attorneys general, who will insist that the Clean Air Act applies to all shipping within the two-hundred-mile limit, so the ships will have to shift to cleaner fuel systems, such as diesel-electric, and scrub what comes out of their stacks if they want to do business on the West Coast.)

Efficiency improvement has already accomplished a lot. From 1975 to 2004, the world increased its "energy intensity" (the amount of energy consumed per dollar of GDP) by 32 percent (44 percent in the United States). There is a great deal more where that came from. We have barely begun to wring the last wasteful joule out of our buildings, vehicles, infrastructure, cities, farms, power lines, ships, armies—the whole apparatus of civilization.

But energy efficiency, crucial as it is, can't replace all the coal-fired plants that have to be shut down, and it can't generate power for the burgeoning energy demand of the growing economies in China, India, Africa, and Latin America. That takes us back to baseload and the choice between coal and nuclear.

● A fourth consideration, along with baseload, footprint, and portfolio, is the role of government. In recent years, environmentalists have largely given up on governments, preferring to work with global and local NGOs, with businesses, and among the grassroots. But infrastructure is one of the things we hire governments to handle, especially energy infrastructure, which requires no end of legislation, bonds, rights of way, regulations, subsidies, research, and public-private contracts with detailed oversight. Energy policy is a matter of such scale, scope, speed, *and* patient follow-through that only a government can embrace it all. You can't get decent grid power without decent government power.

These arguments, plus others I'll come to, have persuaded a surprising number of prominent environmentalists to become pronuclear, some enthusiastically, some grudgingly; some noisily, some quietly. Is there any pattern in who they are and why they espouse nuclear? I think there is.

Start with Jim Lovelock, Gaia's prophet. Gaia theory itself is almost a Green religion, and Lovelock's 1957 invention of the electron capture detector led first to the pesticide measurements behind Rachel Carson's revolutionary *Silent Spring* (1962), then to the detection of environmental PCBs, then to the discovery of atmospheric chlorofluorocarbons and their role in ozone depletion. Lovelock has been pronuclear all his life, ever since his medical research with radioactive isotopes back in the 1940s. He promoted nuclear to my reluctant ears when I first met him in 1985. In a much quoted op-ed in England's *Independent* in 2004, he wrote:

> By all means, let us use the small input from renewables sensibly, but only one immediately available source does not cause global warming and that is nuclear energy. . . . Opposition to nuclear energy is based on irrational fear fed by Hollywood-style fiction, the Green lobbies and the media. These fears are unjustified, and nuclear energy from its start in 1952 has proved to be the safest of all energy sources. . . . I am a Green and I entreat my friends in the movement to drop their wrong-headed objection to nuclear energy. Even if they were right about its dangers, and they are not, its worldwide use as our main source of energy would pose an insignificant threat compared with the dangers of intolerable and lethal heat waves and sea levels rising to drown every coastal city of the world. We have no time to experiment with visionary energy sources; civilisation is in imminent danger and has to use nuclear—the one safe, available, energy source—now or suffer the pain soon to be inflicted by our outraged planet.

Pronuclear public opinion in England went from below 5 percent to over 40 percent. In a reversal of previous policy, the government is planning ten new reactors to replace and add to the nineteen reactors that currently provide 20 percent of England's electricity.

Jesse Ausubel, director of the Program for the Human Environment at Rockefeller University, convened a pioneering conference on climate change back in 1979. He originated the idea of *decarbonization,* noting the two-hundred-year trend of humans using fuels with ever fewer carbon atoms—wood to coal to oil to gas, down to zero carbon with hydrogen and nuclear. In 2007 he published a paper in the *International Journal of Nuclear Governance, Economy and Ecology* in which he declared, "Nuclear energy is green. Renewables are not green." His argument was based on footprint analysis. "As a Green," he wrote, "I care intensely about land-sparing, about leaving land for Nature. . . . Considered in watts per square meter, nuclear has astronomical advantages over its competitors." The solar energy equivalent of a 1-gigawatt nuclear reactor, he projected— with his own variation on Saul Griffith's and Gwyneth Cravens's calculations—would require 150 square kilometers (58 square miles); the wind power equivalent, 770 square kilometers (298 square miles); the corn biofuel equivalent, 2,500 square kilometers (965 square miles).

Patrick Moore, a Canadian who got his PhD in ecology, was a cofounder of Greenpeace in 1971 and became its president in 1977. He left in 1986, declaring that the organization and the environmental movement had become antiscience. Still a strong proponent of forestation, he is best known as a Green spokesman for nuclear power, with a paid role these days as cochair of the Clean and Safe Energy Coalition, founded by the Nuclear Energy Institute, a U.S. lobbying organization.

● The late Anglican Bishop Hugh Montefiore had been a trustee of UK Friends of the Earth for twenty years when he became convinced by climate dangers that as an environmentalist he had to support nuclear power. Told that he could not stay on as a Friends of the Earth trustee if he did that, he resigned, and wrote in a 2004 article,

> The future of the planet is more important than membership of Friends of the Earth. . . . The real reason why the Government has not taken up the nuclear option is because it lacks public acceptance, due to scare stories in the media and the stonewalling opposition of powerful environmental organisations.

Most, if not all, of the objections do not stand up to objective assessment.

Tim Flannery is a world-renowned Australian biologist and conservationist who wrote what is considered one of the best books on climate, *The Weather Makers* (2006). In a 2006 column for a Melbourne newspaper, he posited that

> Over the next two decades, Australians could use nuclear power to replace all our coal-fired power plants. We would then have a power infrastructure like that of France, and in doing so we would have done something great for the world, for whatever risks go with a domestic nuclear power industry are local, while greenhouse gas pollution is global in its impact.

The next year he amended his view to support nuclear in places like China, Europe, and the United States that don't have Australia's renewable energy resources. Australia should keep selling its abundant uranium to those markets, he declared, but should not build its own nuclear plants.

John Holdren, President Obama's science adviser, is an energy and environment heavyweight. He has been an expert on nuclear weapons proliferation, a professor of environmental policy at Harvard, a director of Woods Hole Research Center, and long a coauthor with Paul Ehrlich of ecopolicy texts. His stature gave him cochairmanship of the National Commission on Energy Policy, and he wrote the commission's authoritative 2004 report, *Ending the Energy Stalemate*. It recommended, among other things, that Congress expand support for developing a new generation of nuclear reactors, and it pushed for centralized "interim" storage sites for spent fuel—a detour around the Yucca Mountain roadblock. Holdren told a *New York Times* reporter, "I'm often asked, 'Can you solve the climate problem without nuclear energy?' And I say, 'Yes, you can solve it without nuclear energy.' But it will be easier to solve it with nuclear energy."

Jared Diamond, biologist, conservationist, and author of *Collapse* (2004), had read Holdren's report closely, so when he was asked by an audience member at a San Francisco talk if he "agreed with Stewart Brand in supporting the revival of nuclear," he surprised the audience and me by saying yes: "To deal with our energy problems we need everything available to us, including nuclear power."

James Howard Kunstler, fervent opponent of suburbs, wrote a book in 2004 titled *The Long Emergency*. I'm persuaded by neither his expectation of how peak oil plays out nor his views on the fragility of big cities, but many environmentalists are, and they should note that he ends his "Beyond Oil" chapter with the words, "Nuclear power may be all that stands between what we identify as civilization and its alternative."

There is a category of prominent environmentalist that I predict will increase in coming years—the reluctant tolerators. When they express support of nuclear, they are careful to use sentences too complex to be quote-worthy. Al Gore is one such; he told a Congressional hearing what he usually avoids saying in public, that he is not opposed to nuclear power and expects it will grow somewhat. My old teacher Paul Ehrlich says that climate issues have made him more supportive of nuclear. Bill McKibben, author of *The End of Nature*, penned a friendly review of Gwyneth Cravens's *Power to Save the World* in the environmentalist publication *OnEarth*. "Environmentalists need to understand that times and circumstances change," he wrote, "and they need to rethink priorities. It's not enough for greens to say that nuclear power is risky and comes with consequences; everything comes with consequences." McKibben said he goes along with the IPCC proposal that nuclear "should provide 18 percent of the planet's electricity, up from 16 percent at the moment."

Looking at these Green nuclear proponents—twelve, if you include me—what is the pattern? All but one are obsessed with climate (Patrick Moore is not), and all but four are scientists (Montefiore, Kunstler, Gore, and McKibben are not). With everyone I've encountered who is really immersed in climate issues, the common view of nuclear is "the lesser of two evils" and "take nothing off the table." As for scientists, Gwyneth Cravens reported in her book that they invariably poll high in support of nuclear, ranging from 89 percent among scientists in general up to 95 percent for energy scientists and 100 percent for nuclear and radiation scientists. *(Those who know the most are the least frightened.)*

● And there is a generational element. For a fine piece of journalism about the nuclear debate within the environmental movement, Jason Mark, editor of *Earth Island Journal,* did his research online. As he reported in his 2007 article, "The Fission Division,"

> The anti-nuclear consensus among environmental policy professionals . . . does not extend to the grassroots. A review of some of the most popular green news and opinion Web sites reveals a lively discussion about the merits of expanding nuclear power generation. For example, when Grist.org asked readers, "In the light of the mounting threat of climate change, does nuclear power deserve another look?" 54 percent of respondents voted "Yes." A poll on Treehugger.com showed 59 percent of readers conditionally in favor of atomic energy.
>
> Whenever the issue comes up in green forums, an energetic back-and-forth ensues. During one online discussion, a visitor to a blog hosted by Earthjustice Legal Defense wrote: "I have been an ardent foe of nuclear power generation for over three decades. . . . However, in the last two years I have reversed my position, and now support the building of a new generation of nuclear plants in the USA. The reason is that global warming is such a huge and imminent issue, that I think we must accept the lesser evil of nuclear power generation." When the subject came up on the Web site WorldChanging. com, readers were split roughly 50-50.

A generation gap has emerged. For younger Greens, cold-war nuclear fears are ancient history, and Chernobyl is not part of their personal experience. The threat of climate change is what dominates their world, along with accelerating technology, with which they are comfortable. From the perspective of the young, nuclear is just another technology, to be judged on how well it works, not on antiquated obsessions of their elders.

In 2009 the Director of Greenpeace UK from 2001 to 2007, Stephen Tindale, told the *Guardian* he was now supporting nuclear, and he wasn't alone:

> My change of mind wasn't sudden, but gradual over the past four years. But the key moment when I thought that we needed to be extremely serious was when it was reported that the permafrost in Siberia was melting massively, giving up methane, which is a very serious problem for the world.
>
> It was kind of like a religious conversion. Being anti-nuclear was an essential part of being an environmentalist for a long time but now that I'm talking to a number of environmentalists about this, it's actually quite widespread—this view that nuclear power is not ideal but it's better than climate change

Older environmentalists talk about nuclear power exclusively in terms of what they see as the four great problems that condemn the technology—safety, cost, waste storage, and proliferation. Those four have no form of positive, only degrees of badness, and they are treated as absolutes. If a reactor accident is possible, then nuclear power is impossible; if the capital costs are high, then nuclear power is impossible, and so on. Absolutes are potent. Once something is seen as a capitalized Absolute Evil, it functions as a premise; everything has to exist in relation to your opposition to it.

By contrast, the four considerations I began with—baseload, footprint, portfolio, and government-scale—are logics rather than problems. They are relative rather than absolute, which means they invite thinking in terms of trade-offs and risk balancing. And they are open to the positive, treating nuclear as one potential tool to help head off climate change and end poverty worldwide.

Holding all eight logics and problems in mind simultaneously nets out, for me, to a strong argument for expanding nuclear power. From that perspective, I see the four problems of safety, cost, waste handling, and weapons potential differently than I used to. I've learned to disbelieve much of what I've been told by my fellow environmentalists, and I now think of the four problems the way an engineer does, as *design* problems. Define them, frame them in a way that is solvable, solve the damn things, and once you've got a solution, act on it.

Reactor safety is a problem already solved. In 2008 the world had 443 civilian nuclear reactors boiling up 16 percent of all electricity and keeping a yearly 3 gigatons of carbon dioxide that would have been generated by coal plants out of the atmosphere. Year after year, the industry has had no significant accidents, having learned hard lessons from the three that got away—England's Windscale fire in 1957, the Three Mile Island meltdown in 1979, and the Chernobyl steam explosion in 1986.

"Because a single new accident could destroy the entire nuclear industry worldwide," Tim Flannery argues, "lots of work has gone into minimizing the risk of accidents. As a result, new nuclear technology is relatively safe." Bill McKibben makes a different point: "Nuclear power is a potential safety threat, if something goes wrong. Coal-fired power is guaranteed destruction, filling the atmosphere with planet-heating carbon when it operates the way it's supposed to."

McKibben's pragmatism suggests we reassess our fears in light of real risks closely examined, and that begins with noticing how fear works. An article in *Psychology Today*, "Ten Ways We Get the Odds Wrong," lists some of the elements that play out in nuclear dread:

> We fear spectacular, unlikely events. . . .
>
> We underestimate threats that creep up on us. . . .
>
> Risk arguments cannot be divorced from values. . . .
>
> We love sunlight but fear nuclear power. . . . The word radiation stirs thoughts of nuclear power, X-rays, and danger, so we shudder at the thought of erecting nuclear power plants in our neighborhoods. But every day we're bathed in radiation that has killed many more people than nuclear reactors: sunlight. It's hard for us to grasp the danger because sunlight feels so familiar and natural.
>
> We should fear fear itself. Though the odds of dying in a terror

attack like 9/11 or contracting Ebola are infinitesimal, the effects
of chronic stress caused by constant fear are significant.

● Particularly germane to nuclear is "Risk arguments cannot be divorced from values." Because Hiroshima and the cold war threw everything atomic into the Absolute Evil category, our feelings about nuclear energy are tainted by our revulsion about nuclear weapons. Thus the chemical release of the Bhopal incident in 1984 is treated as far less consequential than the radiation release from Chernobyl, even though over six thousand died from Bhopal versus fifty-six from Chernobyl (forty-seven workers, nine children). Through fear of radiation and expected birth defects, the World Health Organization reported, couples in the Chernobyl area had 250,000 abortions. But no human birth defects have been found to result from Chernobyl—nor, by the way, from Hiroshima. As for disease, the low-dose-radiation expert John Goffman declared that "the number of fatal cancers to come over the years and decades ahead as a direct result of this accident will not be lower than 500,000." In fact it was less than 1 percent of that figure.

"It is roughly estimated that the total number of deaths from cancers caused by Chernobyl may reach 4,000 among the 600,000 people having received the greatest exposures." So says the Chernobyl Forum's report, which was exhaustively researched by seven UN agencies and published in 2006. Its findings are summarized online at the ever-illuminating GreenFacts.org site. About 100,000 of the 600,000 most exposed would die of cancer in any case, if Chernobyl had never happened. For every 100 of the normal cancer victims, four may die earlier because of Chernobyl radiation. Statistically it is a nonevent—epidemiology can't detect it.

The real damage to people in the region, according to the Chernobyl Forum report, is from poverty and mental stress. "The most significant public health impact of Chernobyl has been on mental health," says Luisa Vinton, who headed the forum, in the video *Living with Chernobyl* (2007). "The conclusion we've come to is that fear of radiation is a far more important health threat than radiation itself." The report asserts that what the region around Chernobyl now needs more than anything else is economic stimulus.

● The report also mentions that "the Exclusion Zone has paradoxically become a unique sanctuary for biodiversity." In a fascinating book, *Wormwood Forest: A Natural History of Chernobyl* (2005), Mary Mycio observes that

> It is one of the disaster's paradoxes, but the zone's evacuation put an end to industrialization, deforestation, cultivation, and other human intrusions, making it one of Ukraine's environmentally cleanest regions—except for the radioactivity. But animals don't have dosimeters. . . . The Rhode Island–sized territory has become a fascinating and at times beautiful wilderness teeming with beavers and wolves, deer and lynx, as well as rare birds such as black storks and azure tits.

The rare white-tailed eagle now thrives in the area, and Europe's only remaining native megafauna, the long-endangered European bison and Przewalski's horse, have been introduced successfully. In 1994 two biologists from Texas Tech, Ronald Chesser and Robert Baker, began fifteen years of research on radiation effects in the animals in the Exclusion Zone. In the famously radioactive Red Forest (where all the pines had been killed by fallout, though the birches survived), they started with wild mice.

> We were surprised to find that although each mouse registered unprecedented levels of radiation in its bones and muscles, all the animals seemed physically normal, and many of the females were carrying normal-looking embryos. This was true for pretty much every creature we examined—highly radioactive, but physically normal. It was the first of many revelations.

They did find elevated levels of genetic mutations in voles and reported their results in a 1996 cover story in *Nature*. Right after it was published, they got an automated sequencer, and instead of refining their findings, it completely refuted them—nothing genetically significant was going on in the voles. With anguish, they retracted the *Nature* paper. Robert Baker concludes on his home page that as far as the Chernobyl animals are concerned, "the elimination of human activities such as farming, ranching, hunting and logging are the greatest benefit, and it can be said that the world's worst nuclear power plant disaster is not as destructive to wildlife populations as are normal human activities. Even where the levels of radiation are highest, wildlife abounds."

That is an interesting statement: "The world's worst nuclear power plant disaster is not as destructive to wildlife populations as are normal human activities."

I predict there will be a Chernobyl National Park. It has the perfect ingredients. It is a major historic site. When 330,000 people moved out and wildlife moved in, it became one of Europe's finest natural preserves. The UN Development Programme has called for ecotourism development there. With the exception of a few well-known (and fading) hot spots, radioactivity has dropped to normal background levels. Already reforested, the ghost town of Pripyat, which once housed 50,000, is a poignant reminder of the cost of design folly—the Chernobyl reactors had no containment structures. The site of the reactor itself is a grim monument which, once the final protective shield is in place, may last as long and as evocatively as Stonehenge.

● Several inappropriate Absolute Evils distract rational discussion about nuclear safety. One is cancer. Jim Lovelock, whose degree and early career were in medicine, wrote in 2004, "We must stop fretting over the minute statistical risks of cancer from chemicals or radiation. Nearly one third of us will die of cancer anyway, mainly because we breathe air laden with that all-pervasive carcinogen, oxygen." Cancer researcher and entrepreneur William Haseltine explains another inevitability: "Cancer is a disease of aging. It's going to be hard to prevent cancers, because they are so intrinsically tied to the aging process itself." The surest way to prevent cancer is the traditional method: Die younger of something else.

Radiation acquired its Evil reputation mainly as a legacy of the atomic bomb. Our horror at Hiroshima is transferred, like referred pain, to the only thing we can directly influence, which is nuclear power. For a sense of the legacy effect, compare other dangerous infrastructure-scale energy forms. Gasoline is intensely explosive, killing thousands routinely in auto accidents and elsewhere. Natural gas blows up houses, flows through vulnerable pipelines, and travels around in bomblike ships. Grid electricity is lethal all the way to your light fixture, electrocuting any who touch it carelessly, and setting no end of fires. The great Green hope, hydrogen, has reactivity that weakens pipes, volatility that makes it the most leak-prone of gases, and an ignition point so low that a cellphone can ignite it; and it burns with an invisible flame. Another fond hope, vast quantities of carbon dioxide to be captured at coal-fired plants and piped to underground sequestration, introduces yet another dangerous gas to the energy inventory. In 1986 a dense cloud of CO_2 belched out of Lake Nyos in Cameroon, flowed downhill at 20 miles an hour, and suffocated 1,700 villagers and 3,500 livestock up to 15 miles away. Known all too well to miners as "choke damp," carbon dioxide kills instantly in any concentration greater than 15 percent.

Radiation from nuclear energy has killed not a single American, but of all these energy by-products it is the only one we dread. Nuclear radiation as used in medicine for diagnosis and treatment has saved countless lives, while exposing all the patients to levels of radiation that are many times what is illegal in the nuclear power industry.

● An abiding issue that should not abide much longer concerns the accumulated effects of low-dose radiation. In the 1970s I ran occasional screeds on the subject from Helen Caldicott and the late John Goffman in *CoEvolution Quarterly*, until I read about an experiment comparing the health of lab rats exposed to mild radiation with rats not so exposed. Instead of getting sick, the dosed rats did *better* than the control rats; radiation was good for them. I published no more on the subject.

Decades later, in 2007, I asked Jim Lovelock for his views on low-dose radiation. "The jury is still out, is my answer" he said. "There are two schools. One school says low dosage is good for you. It's called *hormesis*. There's growing evidence for that; it looks quite interesting. The other one says that low dosages are more dangerous. The evidence that impresses me is this: In some parts of the world, in India and Iran, the natural background radiation is huge. The life expectancy of people in those places is the same as anywhere else. There's no evidence of anything nasty going on."

Because background radiation everywhere is considerable, the only way to prove or disprove the standard "linear no-threshold" model, which insists that *all* levels of radiation are harmful to some degree, is to run experiments in a place with no background radiation whatever. That is exactly the plan with a proposed ultra-low-level radiation biology laboratory in the WIPP (Waste Isolation Pilot Plant) nuclear repository deep in a New Mexico salt formation.

Background radiation levels in the salt chambers a half mile down are a tenth of surface levels, and shielding can reduce the radiation to near zero. Comparative experiments with cells, tissue cultures, and transgenic mice highly susceptible to cancer should determine once and for all which of the three theories of low-dose radiation is right. Is it the no-threshold theory? Or, if there is a threshold beneath which radiation damage is negligible, what is the threshold? Or, if the hormesis theory of biopositive effects from low-dose radiation turns out to be correct, what is the most advantageous dose level, and what are the mechanisms of benefit—DNA repair, scavenging of toxic agents, removal of damaged or cancerous cells, or something else? (The hormesis expert T. D. Luckey reportedly finds about 6,000 millirems a year to be the optimum dose level for health.)

At stake are hundreds of billions of dollars. If the standard no-threshold theory is wrong, as most scientists suspect it is, then the U.S. government has been overspending by orders of magnitude on nuclear cleanup, and it will overspend surreally on heading off trace contamination from projects such as the Yucca Mountain repository. The Environmental Protection Agency relies on the no-threshold model to require that nuclear sites do whatever it takes to expose the nearby public to no more than 15 millirems of radiation a year. That is half a mammogram (30 mrem), a fifth of the *range* of normal

U.S. background radiation (from 284 mrem in Connecticut to 364 mrem in Colorado), a sixty-sixth of a CT scan (1,000 mrem; 62 million are done every year in the United States), and an eightieth of one year of smoking a pack and a half of cigarettes a day (1,300 mrem; there are 45 million smokers in the United States). People in Ramsar, Iran, live with a background radiation of 13,000 millirems a year with no apparent health consequences. Astronauts are allowed 25,000 millirems per shuttle mission.

It appears to me that the main public safety issue around nuclear power is what Luisa Vinton and the United Nations agencies found at Chernobyl: "Fear of radiation is a far more important health threat than radiation itself." The lesson of Chernobyl is double: one, be careful; two, be careful what you fear.

✦

That nuclear's high capital cost is the leading complaint from everybody opposing it is due to the prodigious efforts of one man. Amory Lovins, founder and head of Rocky Mountain Institute, has been slicing and dicing the arcana of nuclear-power finance for three decades, to formidable effect.

Lovins is an old friend and colleague. In *CoEvolution Quarterly*, I purveyed his perspective on energy efficiency and praised his books. One time at a Hackers' Conference, I saw him blow away that hard-to-dazzle audience with his discourse on how conventional cars spend most of their energy getting out of their own way, how the nested nuances of his Hypercar design could deliver spectacular energy efficiencies, and how those savings in aggregate would transform the world energy economy. In his writings and talks, Lovins has a pungency of phrasing that rivals the silver-tongued economist John Maynard Keynes. "If," he says offhandedly in an interview, "you build an efficient, diverse, dispersed, renewable electricity system, major failures—whether by accident or malice—become impossible by design rather than inevitable by design."

The Lovins brief against nuclear (or what one opponent calls his "obsessive ongoing vendetta against nuclear") makes the following case. Private capital is the most objective judge and proof of which energy forms are best, and in that arena, nuclear always loses and micropower always wins, and that's why micropower is soaring throughout the world while nuclear is stalled. By micropower Lovins means efficiency, cogeneration (also known as CHP—combined heat and power), small hydro, and renewables such as wind, solar, biomass, and geothermal. "The cheapest, most reliable power is typically produced at or near customers." To Lovins, government intervention to assist nuclear because of climate fears is actually counterproductive because it tilts investment away from the micropower alternatives, which are faster and cheaper to deploy and more effective against climate change. "In no new nuclear project around the world is there a penny of private capital at risk."

In 1986 a TV interviewer asked Lovins about the future of nuclear power. "There isn't one," he replied. "No more will be built. The only question is whether the plants already operating will continue to operate during their lifetime, or whether they'll be shut down prematurely." At a debate in 2007 he declared, "Nuclear is dying of an incurable attack of market forces despite what the industry wants you to believe."

• In 2007, as Lovins was speaking, the following was taking place. Around the world, thirty-one reactors were under construction. In the United States, the Nuclear Regulatory Commission was bracing for an expected seventeen new nuclear plant applications (they were still on track in 2009), while Canada was preparing for new reactors in Alberta and New Brunswick (that one is to sell part of its power to Maine, which shut down its only nuclear plant in 1997). In Europe one nation after another was reversing course on nuclear. Green Finland started construction on a huge, 1.6-gigawatt plant, then added another. Green Germany quietly decided to keep its seventeen reactors working rather than shut them down as previously announced. Green Sweden did the same with its ten reactors while expanding their capacity. Green Belgium decided likewise with its seven reactors. Italy had shut down its four reactors in 1987, after Chernobyl, and soon became Europe's largest energy importer, mostly from nuclear France; in 2007 it began a program to build four new reactors. Gordon Brown's Labor government in England made the decision to upgrade all of its nineteen reactors and build new ones. Ireland, Norway, and Poland were planning their first reactors. France, already getting 80 percent of its electricity from nuclear, began building new reactors and expanding its comprehensive nuclear operations to serve a global market through its $7 billion public corporation, AREVA. Russia set off on the same path, planning to double its number of reactors, currently thirty-one, and sell to an international market with its new $8-billion state-owned holding company, Atomenergoprom.

In the developing world, Russia's interested potential customers include "Vietnam, Malaysia, Egypt, Namibia, Morocco, South Africa, Algeria, Brazil, Chile and Argentina," according to the *New York Times*. The article added that "Western investors, including such heavyweights as Citigroup, are seeking ways to bet on the Russian nuclear power industry." Now building nukes for the first time are Albania, Belarus, Turkey, Iran, the United Arab Emirates (two 1.6 gigawatt reactors under way), Burma, Thailand, Vietnam, and Bangladesh; considering doing so are Syria, Israel, Jordan, Saudi Arabia, Bahrain, Kuwait, Oman, Qatar, Algeria, Ghana, Nigeria, Kazakhstan, Indonesia, and the Philippines; and expanding on the nukes they already have are Ukraine, Hungary, Bulgaria, Armenia, Argentina, Mexico, Brazil, South Africa, Pakistan, South Korea, and Taiwan. (You can keep track of the count at Wikipedia's entry, "Nuclear energy policy.")

India, with seventeen reactors now providing 3 gigawatts of power, plans to expand the capacity tenfold to 30 gigawatts.

After ordering two reactors from Westinghouse, China signed a $12 billion deal with AREVA—said to be the largest in the history of the industry—and is aiming to grow its nuclear capacity to 70 gigawatts by 2020. It has eleven reactors working, five under construction, thirty planned, and eighty-six proposed.

Japan, the largest user of nuclear power after the United States and France, with fifty-five reactors in operation, is planning eleven new reactors by 2017. The government wants to transform its electricity mix from 30 percent nuclear now to 60 percent nuclear by 2050.

In early 2009, in *Ambio* magazine, Amory Lovins declared: "Nuclear power is continuing its decades-long collapse in the global marketplace because it's grossly uncompetitive, unneeded, and obsolete."

How can someone so smart be so wrong about a subject he knows so well? It turns out that his arguments against the economics of nuclear power work only within the narrow commercial boundaries he defines, which increasingly no longer apply, and he focuses mainly on the United States. His reasoning has no traction in relatively dirigiste economies like France, Japan, and most developing countries, especially China and India; if those governments want nukes, they build nukes. More important, the loom of climate change has altered everybody's perspective on costs and risks.

● The problem is not that nuclear is expensive. The problem is that coal is cheap.

"Nuclear is dying of an incurable attack of market forces," said Lovins. It was market forces that gave us coal's dominance—40 percent of the world's electricity comes from coal (and 20 percent from gas, 16 percent from nuclear, 16 percent from hydro, 7 percent from oil, 2 percent from renewables). If only market forces rule, coal will continue to beat everything else, and the world's goose is cooked.

That realization motivates governments even in strongly capitalist societies like the United States and Britain. By government fiat, coal will be made expensive—through carbon taxes, cap-and-trade markets, requirements for CCS (carbon capture and sequestration), and mandates. In the competition to provide baseload power, as the cost of coal-fired electricity goes up, nuclear will take the lead as the most cost-effective alternative. (The economics of wind, hydro, cogeneration, solar, and geothermal also gain in relation to a properly crippled coal.) The fate of gas-fired plants, including cogeneration, will depend on the local price of natural gas, which in most places is rising precipitously. And no nation that has been dependent on Russia for natural gas wants to remain so, because it gets used as a political weapon.

The sticker shock for nuclear comes right at the start. Like the other major Green power sources—hydroelectric dams and large wind farms—nuclear plants require massive up-front capitalization; but once the facility is built, operating costs are low compared to fuel-intensive coal and gas plants, accounting for only one quarter to one third of the plant's lifetime expense.

Wind is serious infrastructure, and growing fast—94 gigawatts capacity worldwide in 2007 (nuclear was 365 gigawatts; coal, 1,393 gigawatts). Expanding wind power always means expanding long distance power lines. In Denmark and Germany, which led the European wind revolution, new construction has tapered off due to rising costs and scarcity of appropriate sites. Some are offended by the sight of massive wind farms churning away (Jim Lovelock and Robert Kennedy Jr. come to mind), but personally I find them thrilling. Unfortunately, wind power remains limited by intermittency to about 20 percent capacity (so that 94 gigawatts is four-fifths illusory), while nuclear plants run at over 90 percent capacity these days; and there still is no proven storage technology that would make wind a baseload provider. (Potential massive energy storage techniques that are being explored include flywheels, hot liquid, compressed air, and better batteries or distributed batteries in everybody's plug-in hybrid car.)

Many bird-loving environmentalists fight wind farms because of the harm to birds. My favorite Green publication, *High Country News*, ran the following list without comment: "Annual bird kill in the US: wind turbines, 28,500; buildings, 550 million; power lines, 130 million; cats, 100 million; cars, 80 million; pesticides, 67 million."

• Solar so far is a bit player in electricity generation—10 gigawatts of capacity total in the world in 2007, but with solar's 14 percent capacity factor, that's only 1.4 gigawatts operational, less than one large nuclear reactor. I fondly remember the 1970s solar boom, which ended the moment Ronald Reagan became president and canceled Jimmy Carter's solar tax credits. I had the delicious opportunity to interview Ted Turner in front of the 2007 Solar Power Conference, which had twice the attendance of the previous year, 12,500 people. Among them were only a few of the old solar guard such as John Schaeffer, whose solar mail-order catalog, *Real Goods*, has freed thousands from the grid since 1978. Schaeffer and I were surrounded by suits, most of them apparently avid to save the world *and* become rich from a technology they expected to be heavily subsidized any day. That's why they wanted to hear from rich, Green, Ted Turner. Turner went cheerfully off topic to inveigh against the Iraq War, noting that the enemy motivated their troops by promising them forty virgins when they die. "If we had incentives like that for solar power," Turner cracked, "we'd have it made." (I couldn't resist asking Turner about nuclear; he said, "I would rather have a nuclear plant than a coal-burning plant.")

Boundless ingenuity and capital are being focused on solar, with new materials and new configurations—nano-this and concentrated that; someday, maybe, energy-producing solar paint. The most efficient so far is concentrated solar thermal, in which focused mirrors heat a fluid to run turbines, and some of the heat can be stored in steam accumulators when the sun goes behind a cloud or sets for the night. With enough breakthroughs, sustained over decades, solar could and should become a leading source of electricity. Hasten that day! Meanwhile, the finance-management firm AllianceBernstein opines that "the solar industry has benefited recently from spectacular policy-driven growth. . . . We think that the speculative interest is becoming bubble-like, particularly given that industry fundamentals depend on government subsidies and political support."

My wife, Ryan, and I use plenty of solar with no need for incentives. We've had solar panels trickling maintenance-free power into the battery bank on our tugboat for twenty-six years. The electric fence that keeps cattle out of the oak savannah we're restoring runs on solar, and so does the automatic gate in the fence. Best of all is the solar water heater for our lap pool. We get an 80°F pool for seven months of the year with no use of propane, and the panels work in reverse on hot days, radiating excess heat to the night sky. Solar works well at the individual level; wind does not. Wind works well at the infrastructure level; solar, so far, does not.

• A standard objection to nuclear, following Amory Lovins's lead, is that it is grossly oversubsidized: It could never compete in a fair market, it steals subsidies that solar should get, etc. A 2007 study compared U.S. government energy incentives over the past fifty years for oil, coal, gas, hydro, nuclear, wind, solar, and geothermal, with particular focus on the period from 1994 to 2003. The incentives tallied included direct subsidies, research, tax relief, regulation, government services, and market intervention. Among the conclusions were these:

> The general perception that the oil industry has been the major beneficiary of Federal subsidies is correct, with this source receiving nearly half of all subsidy support. . . . The perception that renewable energy has been short-changed at the expense of other energy sources is not correct. . . . Coal and nuclear technologies have been underfunded, while solar technologies such as photovoltaics, solar thermal, and wind have been well funded.

(Details, with illuminating graphs, may be found in Roger H. Bezdek and Robert M. Wendling, "A Half Century of United States Federal Government Energy Incentives: Value, Distribution, and Policy Implications," *International Journal of Global Energy Issues*, vol. 27 [2007], no. 1.) The study leaves out the copious subsidies at the state level for renewable energy.

"Clean coal" is supposed to be a low-carbon alternative to nuclear for baseload power once we have carbon capture and sequestration (CCS). As of 2009, there was not a single demonstration plant using CCS, though China is working on one. Even industry proponents of CCS don't expect the first commercial application before 2030 or broad use before 2050. The volume of CO_2 that has to be dealt with is overwhelming. A 1-gigawatt coal-fired plant burns eighty rail cars of coal a day, each car carrying 100 tons of coal. The resulting CO_2 weighs 2.4 times as much as the coal, so that plant will produce over 19,000 tons of carbon dioxide a day—7 million tons a year. It has to be separated, compressed to a liquid, piped to a suitable location, pumped down half a mile, and sealed off permanently. Good luck with that in the developing world. A Swede named Anders Hansson did the math on the billions of tons of carbon dioxide that would have to be buried to make a difference and concluded, "Carbon dioxide would be the world's largest transported good." *Washington Post* cartoonist Tom Toles has it right: The best carbon sequestration technique is to leave the coal in the ground.

You may have heard this one: "It's no good building new reactors because we're running out of uranium." In fact we're not, and it wouldn't matter if we did. Known uranium deposits cover a hundred years of current-level use. The rising price of uranium is driving new exploration, with finds turning up all over the world. The price rise also makes reprocessing of spent fuel more attractive, and that leverages existing uranium manyfold. The leading substitute for uranium, thorium, is three times as abundant, and it can't melt down, it's useless for weapons, and it generates little waste. Meanwhile, new reactors are increasingly efficient, and breeder models create more fuel than they use. I'll come back to the new designs in a minute.

• We Greens are not economists. When someone needs fiscal advice, they don't usually hire an environmentalist, because they know we don't really know about money and we don't really care about money. Our agenda is to protect the natural environment, not taxpayers or ratepayers. We're perfectly happy requiring such impediments as environmental impact reports, which add horrendous costs and delays to projects, and our arguments for protecting one endangered species or another do not include cost analysis. How much is a condor worth? Don't even ask.

Occasionally we'll invent useful economic instruments like debt-for-nature swaps, and many of our organizations are well run financially, but money issues are customarily employed by Greens strictly as a weapon. "If you want to kill a power project," advises one activist in *Orion* magazine, "focus on economics." I recall the futurist Herman Kahn talking about the fight over the Trans-Alaska Pipeline in the 1970s: "The Greens began by complaining about the design of the pipeline, and they were right, it did have flaws. But once those were fixed, the Greens put all their effort into forcing delays and extravagances that raised the cost of the system. Their final argument was that the pipeline should not be built because it was too expensive. That's like a kid murdering his parents and then asking for mercy from the court because he's an orphan."

Amory Lovins does have economic expertise. He deploys cost analysis brilliantly in the service of energy efficiency and conservation. His Rocky Mountain Institute well earns its $9 million a year helping corporate clients and the military make their energy practices both frugal for themselves and good for the environment. But when Lovins aims his well-honed techniques at nuclear power, I think his private-sector bias gets in the way. He has said, "I admire those who try to reform public policy, but

I don't spend much time doing that myself. In a tripolar world of business, civil society, and government, why would you want to focus on the least effective of that triad?"

With global warming, the game has changed. Market forces cannot limit greenhouse gases. Governments have to take the lead. What they deem the atmosphere requires will be the prime driver of the economics of energy.

✦

A common refrain against nuclear goes, "Solving the problem of waste storage is so difficult, not a single geological repository for nuclear waste is operating anywhere in the world."

But there is one in the United States, in operation burying radioactive stuff since 1999. WIPP, the Waste Isolation Pilot Plant in New Mexico, got through its safety reviews thanks largely to Rip Anderson, the scientist at Sandia Labs who guided Gwyneth Cravens through the nuclear world for her book. For political reasons, WIPP was assigned only waste (some low-level, some high-level) from U.S. military activities, while the high-level waste from the civilian energy program was supposed to go to Yucca Mountain in Nevada. Having studied every kind of repository in detail, Anderson declares:

> From a technical point of view, the best place on dry land to store all nuclear waste—wherever it comes from—is at WIPP. We've proven that every way you can think of. We have trace-ability and transparency. Geologically and hydrologically, it's the safest. There's room for it, and more panels can be mined out of the salt bed whenever we want. It's only politics and bureaucracy that stand in the way.

WIPP is a salt formation, and Yucca Mountain is a dry desert ridge in a huge military reservation, but that's not why one is now working and the other isn't. New Mexico is familiar with nuclear technology, through long experience with Los Alamos and Sandia, where nuclear weapons were designed and built. Nevada is where the bombs were tested. Nevadans miss no opportunity to get in a fight with the federal government, which owns 86 percent of the state's land. Nuclear waste in Nevada? "No, God damn it! Take your garbage somewhere else!" Antinuclear environmentalists restate that sentiment in terms of dangers of transport, insufficient testing on groundwater flow, effects of heat on rock, impossibility of defining what will happen over 10,000 years, etc.

Jim Lovelock is baffled by the fuss:

> Consider the Yucca Mountain nuclear-waste depository in Nevada: It cost a fortune to construct, and we need it about as much as we need a facility for imprisoning dangerous extraterrestrials. We must stop living in a sci-fi world. In the real world, high-level nuclear waste from 40 years of energy production in the U.K. and France is stored as chunks of glass packed in stainless steel containers and buried a few meters underground. Sandy [Lovelock] and I stood on all the French high-level nuclear waste at La Hague in Normandy. The radiation level on my own monitor was only 0.25 microsieverts an hour, which is about 20 times less than you'd find in any long-distance passenger plane.

In Finland, construction is going ahead on a deep geological repository for high-level wastes at Eurajoki, where the project found a welcome because the local population had become comfortable with nuclear power: It has been generated in their neighborhood for thirty years. The repository should open for business in 2020. Meanwhile, Finland's neighbor Russia wants to become a nuclear fuel handler for the world—mining, processing, shipping, and reprocessing fuel, and then storing the waste.

The method of dealing with nuclear waste in the United States that emerged, de facto, in the absence of a national repository is proving so practical that a new solution is based on it. U.S. reactor operators now put their spent fuel in pools to cool off for a few years, then pack it in dry cask storage out behind the parking lot. So it is for all 121 reactor sites in thirty-nine states. Since the casks are designed for transport (there's a thrilling video made at Sandia of trains, planes, and trucks ramming the casks without release of spent fuel), the idea now being legislated—with support of the

Obama administration—to cart them to a few well-guarded "interim" sites where they can be parked for decades while the nation decides whether it wants to recycle the material or bury it. Thus the United States would adopt Canada's "adaptive phased management" solution. To my eye, the emerging rule is: Plan short and option long; take the actions in the near term that preserve the most choices for the long term.

That approach was taken with funding for waste storage. American purchasers of nuclear-generated electricity have been paying for eventual storage facilities for years. Some $28 billion has accumulated. What's in short supply now is political decisiveness.

• I used to be certain that reprocessing spent fuel is the right thing to do. The once-through "waste" we bury has 95 percent of its energy still in it. If all the existing spent fuel in the United States were reprocessed, our whole nuclear fleet could run for seven years on nothing else. The eventual volume of unreusable waste after reprocessing is a fraction of the usual waste—some say a fourth, others a tenth—and the radioactive material does not remain dangerous for nearly as long. Because weapons-capable plutonium is generated in the current forms of reprocessing, Presidents Gerald Ford and Jimmy Carter shut down the U.S. operation in the hope that the rest of the world would follow suit. It didn't work. Routine in France, Russia, England, Germany, and Japan, reprocessing is being considered in many nations with ambitious nuclear-energy programs. These days, the United States itself is undertaking to build reprocessing plants in Idaho, New Mexico, and South Carolina.

An article in the engineering magazine *IEEE Spectrum* on France's exemplary reprocessing operation is what gives me some second thoughts. The facility at La Hague reprocesses 1,700 tons of spent fuel a year, with an impeccable safety record. But even in France, with its primary reliance on nuclear for electricity, the enormous cost of reprocessing makes it economic only if the price of uranium is very high, or if the nation is developing fast breeder reactors. While the eventual waste from reprocessed fuel is relatively short-lived, it is much "hotter" and trickier to handle than the once-through spent fuel.

And then there's the weapons issue. That's the one that keeps Al Gore hesitant about nuclear. He has said:

> For eight years in the White House, every weapons-proliferation problem we dealt with was connected to a civilian reactor program. And if we ever got to the point where we wanted to use nuclear reactors to back out of a lot of coal—which is the real issue: coal—then we'd have to put them in so many places, we'd run that proliferation risk right off the reasonability scale. And we'd run short of uranium, unless they went to a breeder cycle or something like it, which would increase the risk of weapons-grade material being available.

Current reprocessing techniques produce plutonium and enriched uranium, which are potential bomb materials. There are political and technical workarounds, though.

● First, it should be said that nuclear energy has done more to eliminate existing nuclear weapons from the world than any other activity. Megatons to Megawatts (I guess they decided against Megadeaths to Megawatts) is the name of a joint U.S.–Russia program to convert warheads into fuel. It began in 1994, and currently 10 percent of the electricity Americans use comes from Russian missiles and bombs. The goal is to convert twenty thousand nuclear warheads into fuel by 2013; that's enough energy to run the whole U.S. nuclear fleet for two years. Two processes are involved. One "downblends" weapons-grade highly enriched (95 percent) uranium to low-enriched (5 percent) uranium for reactor use. The other converts plutonium into a "mixed oxide" (MOX) that also works as a fuel. In the next few years, the United States will supplement Russian efforts by commencing to forge our own nuclear swords into plowshares in Tennessee and at a new facility in South Carolina. This whole astonishing program has occurred with scant media attention and zero public hoopla. (Freeman Dyson tells me that's good: "Important moves toward disarmament always go better when they are not reported in the media. Historic examples are

Nixon's getting rid of U.S. biological weapons and George Bush Senior getting rid of most of the U.S. tactical nukes. Both these big disarmament moves were done quietly so that they did not become political issues.")

No other weapons system creates so much civilian value when dismantled. The sundry calls from around the world and across the political spectrum to eventually eliminate *everybody's* nuclear weapons inventory and apparatus can draw on this strong additional incentive.

There is logic for and against the argument that expanding nuclear energy expands the possibility of nuclear weapons proliferation. The observation that "more nuclear means more nuclear, period" has bracing clarity. A worldwide nuclear energy renaissance involves massive expansion of nuclear skills, nuclear equipment, and nuclear materials—a whole nuclear economy. Nuclear weapons programs should have no problem quietly expanding in the shadow of all that, right?

Among the critiques of this energy-to-weapons sequence, perhaps the strongest argument is that it hasn't occurred that way in history. Israel, India, South Africa, and North Korea secretly developed their weapons capability from research reactors, not from an energy program. Of the thirty-one nations currently using nuclear power, only seven have nuclear weapons (the United States, Russia, England, France, China, India, and Pakistan), and in each case the weapons program came first. North Korea and Israel have weapons but no nuclear power (though both have energy projects under way).

What I hear from people in the intelligence world is that antiproliferation efforts are going surprisingly well, mainly because international cooperation is inclusive and intense. It doesn't matter if a particular American administration is in bad odor; every nation has its own strong motivation to have no nuclear weapons capability in non-state-actor hands. The full argument against the nuclear terrorism threat is worth hearing from John Robb, who runs the brilliant blog *Global Guerillas* and wrote *Brave New War* (2007):

> Though most of the worry over WMDs [weapons of mass destruction] has focused on nuclear weapons, those aren't the real long-term problem. Not only is the vast manufacturing capability of a nation-state required to produce the basic nuclear materials, but those materials are difficult to manipulate, transport, and turn into

weapons. Nor is it easy to assemble a nuke from parts bought on the black market; if it were, nation-states like Iran, which have far more resources at their disposal than terrorist groups do, would be doing just that instead of resorting to internal production.

It's also unlikely that a state would give terrorists a nuclear weapon. Sovereignty and national prestige are tightly connected to the production of nukes. Sharing them with terrorists would grant immense power to a group outside the state's control—the equivalent of giving Osama bin Laden the keys to the presidential palace. If that isn't deterrent enough, the likelihood of retaliation is, since states, unlike terrorist groups, have targets that can be destroyed. The result of a nuclear explosion in Moscow or New York would very probably be the annihilation of the country that manufactured the bomb, once its identity was determined—as it surely would be, since no plot of that size can remain secret for long.

Even in the very unlikely case that a nuclear weapon did end up in terrorist hands, it would be a single horrible incident, rather than an ongoing threat. The same is true of dirty bombs, which disperse radioactive material through conventional explosives. No, the real long-term danger from small groups is the use of biotechnology to build weapons of mass destruction. In contrast with nuclear technology, biotech's knowledge and tools are already widely dispersed—and their power is increasing exponentially.

(I'll get to Robb's final point on biotech in a later chapter. I should add that his view of Iran is that, whatever its ambitions for nuclear weapons, it does actually need nuclear energy, and that gives the world bargaining leverage on exactly how Iran proceeds with nuclear technology.)

• The many minds over many decades that have focused on preventing further nuclear proliferation have converged in the last few years on a single, all-encompassing strategy—an international fuel bank. Rid the world of the creation of weapons-grade plutonium and uranium and of hidden pockets of those materials by closely managing the world market in nuclear fuels and spent fuels so that unsupervised enrichment and reprocessing cannot occur. There is no explosive capability in nuclear fuel or spent fuel by themselves: They're far too dilute. Many nations would welcome the outsourcing of their fuel reprocessing, because it is so expensive and tricky, and it would be a great relief for them to have their nuclear waste go away to be treated as someone else's resource or service instead of lingering as a politically fraught local storage problem. The nation would basically rent nuclear fuel from a trusted international facility that comes around like the milkman to drop off milk and pick up empty bottles.

That is the idea behind GNEP—the Global Nuclear Energy Partnership—which could wind up being the one lasting accomplishment of the George W. Bush administration. Put forward in 2006, GNEP was blasted by antinuclear organizations and criticized by the National Academy of Sciences; but the partnership was quickly joined by eighteen nations, including France, Japan, Canada, and Britain, and it got support from the head of the International Atomic Energy Agency, crucial for program oversight. Part of the GNEP scheme is to develop new proliferation-resistant reprocessing techniques and new reactor designs such as the sodium-cooled fast reactor, which breeds its own fuel, reducing traffic in nuclear fuel as well as reducing spent-fuel mass and longevity. Whether or not the GNEP prevails, it is just the most ambitious of several multinational fuel-cycle initiatives being proposed and funded. One or a combination of them is likely to be operative by the 2010s. President Obama announced his support of a nuclear fuel bank program before a global audience in Prague in early 2009.

● One reason the French were able to build a fleet of fifty-six reactors providing nearly all of the nation's electricity in just twenty years was an efficient licensing process that took four years instead of the twelve years that became standard in the United States. As a result, France has the cleanest air in Europe, the lowest electrical bills, and a $4 billion export business selling energy to all its neighbors, including Green Germany and nuclear Britain (2 gigawatts a year flows west under the English Channel). France shut down its last coal-fired plant in 2004. It emits 70 percent less carbon dioxide per capita than the United States.

Lessons learned. The U.S. Nuclear Regulatory Commission has adopted the French approach, with design standardization and a licensing process based on the model that emerged in Taiwan, Japan, and South Korea. Reactor manufacturers can get their designs *pre*approved. One next-generation reactor, from Westinghouse, has already been approved, and two others, from AREVA and GE, are in process. Many reactor-site approvals already exist, left over from the 1970s boom. Once a utility decides to build, it applies for a single combined construction and operating license, and the NRC has three years to grant or deny the license. Get the data, hear the critiques (including those from environmentalists), make adjustments, *decide*, and move on. With standardized designs and standardized parts, construction of large new plants should be completed in four years.

This is the kind of mobilization that is needed to deal with climate change. It should be applied to every activity affecting greenhouse gases, including renewable-energy technologies, transportation, agricultural practices, and city design.

Seeking that kind of acceleration, in 2005 Congress passed the Energy Policy Act with incentives for wind, solar, geothermal, biofuels, wave and tidal power, clean coal, efficiency and conservation . . . and nuclear. As the *Economist* summarized it, the act

> offers four different types of subsidies for new reactors. First, it grants up to $2 billion in insurance against regulatory delays and lawsuits to the first six reactors to receive licences and start construction. Second, it extends an older law limiting a utility's liability to $10 billion in the event of a nuclear accident. Third, it provides a tax credit of 1.8 cents per kilowatt hour for the first 6,000 megawatts generated by new plants. Fourth, and most importantly, it offers guarantees for an indeterminate amount of loans to fund new nuclear reactors and other types of power plant using "innovative" technology.

The act includes funds for research on what are called generation IV reactors. (The reactors currently operating in the United States are generation II, and the ones now being constructed are referred to as generation III-plus—with better built-in safety and efficiency.) Generation IV designs, which are not expected to be commercial before the 2030s, aim for maximum all-around sustainability through lower construction and operating costs, superhigh fuel efficiency, greatly reduced waste with shorter-term radioactivity, and high temperatures capable of generating hydrogen or desalinating water. The goal is to solve permanently every one of the four major problems with current reactors—safety, cost, waste, and proliferation. A consortium of ten nations and the European Union is carrying out the research.

James Hansen suggests that the development of generation IV reactors could be sped up by applying some of the $28 billion already collected from the nuclear industry for waste storage: "This fund should be used to develop fast reactors that consume nuclear waste, and thorium reactors to prevent the creation of new long-lived nuclear waste." The reactors he's referring to are the integral fast reactor, which, he says, "can burn existing nuclear waste and surplus weapons-grade uranium and plutonium, making electrical power in the process," and the liquid-fluoride thorium reactor. Full-scale working reactors, Hansen proposes, could be developed first in China, India, or South Korea, where the need is greatest, with American or European technical support. Deployed in the United States, the waste-burning reactors would make Yucca Mountain irrelevant.

Freeman Dyson comments: "Next-generation reactors could make a big difference. I like especially Lowell Wood's scheme for building a thorium breeder that runs for fifty years without refueling. It is buried deep underground, and after the thorium is burned, it stays in the ground and is never touched." Details on the design may be found in a 2008 paper in *Progress in Nuclear Energy*. The five authors conclude with the reactor's advantages: "No more mining, no more enrichment operations, zero spent-fuel handling, no reprocessing or waste storage facilities, and the reactor vessel is the (robust) burial cask."

• The new nukes with the greatest potential appeal for environmentalists are microreactors. They speak directly to Amory Lovins's call for distributed micropower ("The cheapest, most reliable power is typically produced at or near customers"). They have the advantage of capital costs and construction times a fraction of what is required for the standard gigawatt-plus big nuclear plants, and the development time for new designs also is drastically shorter.

Right now Russia is building 35-megawatt reactors that float on barges, for use starting in 2010 along the nation's newly navigable Arctic coast, and Russia's developing-world clientele is expected to buy some for remote coastal villages. In Japan, Toshiba has invented a 10-to-50-megawatt "nuclear battery" the company calls 4S—for "super safe, small, and simple"—expected to be ready around 2015. Not to be outdone in the acronym department, the Lawrence Livermore Lab in California has a 20-megawatt reactor design it calls SSTAR—"small, sealed, transportable autonomous reactor." The idea with these small units is that they can be trucked in and planted in the ground, and no fuel goes in, no waste comes out—they're simply replaced decades later.

A company in New Mexico called Hyperion is building 25-megawatt reactors using a uranium hydride–based design from the Los Alamos National Laboratory; the company claims it has a hundred firm orders at $25 million per reactor. A company in Oregon, NuScale, has a 40-megawatt light-water reactor design it says can be in operation by 2015. Then there's the pebble-bed modular reactor being developed in South Africa. Meltdown-proof, it also can operate underground, at a scale of 100 megawatts.

✦

To my mind, the Green path forward begins with environmentalists realizing that nuclear power will grow no matter what we do. Our customary opposition would make it grow badly—slowly, expensively, unsystemically, and with dangerously poor overall coordination. But if we encourage it in the right way, nuclear energy growing well would mean that it minimizes humanity's carbon-loading of the atmosphere; that it collaborates well with other carbon-free or superefficient energy forms; that it helps generate other Green services such as desalination or hydrogen; that its uranium and thorium make minimum lasting mess on their way from the ground and back to it; that it helps eliminate nuclear weapons; that it securely energizes cities and thereby helps reduce world poverty; and, if something better comes along, that it gracefully gives way to its replacement.

Glow-in-the-dark Greens might push especially hard for first-rate microreactors and help them to find customers. We might insist that the next version of the Kyoto Protocol does not repeat the disgrace of denying carbon credits for nuclear power. We might lobby for open-sourcing the various reactor designs, for public debugging and trust. We might encourage investigating nuclear propulsion on commercial ships, the current source of 4 percent of greenhouse gas emissions—double the amount generated by airplane traffic. And as we promote plug-in cars, we might note that they reduce carbon emissions only if their electricity comes from Green sources such as nuclear, wind, hydro, and solar.

If hydrogen becomes a practical fuel, we might support high-temperature generation IV reactors that specialize for hydrogen. The same machines could drive desalination, the most energy-intensive source of water and also the least naturally disruptive way to provide water for our increasingly coastal populations. Because coal is now understood to be the long-term systemic horror we once thought nuclear was, we should closely monitor the real-world effectiveness of coal disincentives and make sure they work. To encourage public confidence and familiarity with nuclear, we can follow Sweden's and France's example by opening all reactors to public tours. (A third of all Swedes have toured a nuclear plant, which helps to explain why 80 percent of the population supports their continued use.)

For our fellow environmentalists still queasy about nuclear, we might quote Al Gore's mentor Roger Revelle, who sponsored the atmospheric carbon dioxide studies that first exposed the inconvenient truth about climate change. Revelle regarded nuclear as "much more benign" than other energy sources. He said, "What we ought to do is imitate the French and Japanese. They haven't got any phobias about it." We can even invoke the Sierra Club, which pushed for nuclear power in the late 1960s and early 1970s as preferable to hydroelectric dams. They were right.

• The atmosphere responds to the aggregate of all human activities. What the United States does about nuclear is not the main event.

The squatters' rights organization in South Africa, Abahlali baseMjondolo, has declared: "Electricity is not a luxury. It is a basic right. It is essential for children to do their homework; for safe cooking and heating; for people to charge phones, to be able to participate in the national debate through electronic communication (TV discussion programmes, email, etc); for lighting to keep women safe and, most of all, to stop the fires that terrorise us."

About half of India has no grid electricity. What power there is comes from diesel fuel trucked around to local generators. A *New York Times* article told of the fate of one village:

> Chakai Haat once had power at least a few hours each day, and it changed the rhythm of life. Petty thefts dropped because the village was lighted up. The government installed wells to irrigate the fields. Rice mills opened, offering jobs.
>
> The boon did not last long. Strong rains knocked down the power lines. The rice mills closed. Darkness swathed the village once more.

Five out of six people live in the developing world—about 5.7 billion in 2010. One way or another, the world's poor will get grid electricity. Where that electricity comes from will determine what happens with the climate.

Footnotes for this chapter may be found at **www.sbnotes.com**.

· 5 ·
Green Genes

A truly extraordinary variety of alternatives to the chemical control of insects is available. Some are already in use and have achieved brilliant success. Others are in the stage of laboratory testing. Still others are little more than ideas in the minds of imaginative scientists, waiting for the opportunity to put them to the test. All have this in common: they are biological solutions, based on understanding of the living organisms they seek to control, and of the whole fabric of life to which these organisms belong. Specialists representing various areas of the vast field of biology are contributing—entomologists, pathologists, geneticists, physiologists, biochemists, ecologists—all pouring their knowledge and their creative inspirations into the formation of a new science of biotic controls.

—Rachel Carson, *Silent Spring*, 1962

I daresay the environmental movement has done more harm with its opposition to genetic engineering than with any other thing we've been wrong about. We've starved people, hindered science, hurt the natural environment, and denied our own practitioners a crucial tool. In defense of a bizarre idea of what is "natural," we reject the very thing Rachel Carson encouraged us to pursue—the new science of biotic controls. We make ourselves look as conspicuously irrational as those who espouse "intelligent design" or ban stem-cell research, and we teach that irrationality to the public and to decision makers.

We also repel the scientists whose help we most need to develop a deeply sustainable agriculture: the agronomists, ecologists, microbiologists, and geneticists who are fulfilling Rachel Carson's dream.

● When genetic engineering first came along in the 1970s as recombinant DNA research, I was surprised by the hysteria it inspired. From my biology background, I knew that genes have always been intensely fungible, especially in microbes. We weren't creating a new technology so much as joining an old one, using the very techniques that microbes have employed for 3.5 billion years. Asking around over the years, I've found that professional biologists are universally unalarmed about genetic engineering. Most are adopting it in their own work because it is transforming every one of the biological disciplines.

Activists who oppose genetic engineering in crops give detailed biological arguments about the damage they're sure it will cause, but it's oddly context-free biology. Superweeds! Allergies! Gene flow! Transgenic novelty! Surprises! While the public is saying "Oh my God!" biologists are saying, "So?" All that stuff is normal, in the wild and in agriculture. In the wild, you study it; in agriculture and restoration, you tweak it. Genetic engineering offers some nifty new precision, speed, and reach, but that's all that is new. As with nuclear, those who know the most are the least frightened.

What I'll attempt to do here is fill in some of the biological context that makes genetic engineering unthreatening good news, especially for environmentalists.

A note on usage. Except in quotations, I'll employ the emerging abbreviation GE instead of the still common GM. GM usually stands for "genetic modification" or sometimes "genetic manipulation," but the term is too general. All of evolution, all of agriculture, and all of selective breeding of any kind is genetic modification and always has been. GE specifies genetic *engineering*, a finer-grained practice. Instead of selecting for traits, which breeders do, GE identifies the genes behind the traits and works selectively and directly with those genes. Breeding is usually stuck with the traits and genes within the existing genome of an organism and of species that will hybridize with it. GE can reach out for desired traits and genes in more distant organisms. A synonym for *genetically engineered* is *transgenic*.

There lies the horror, apparently. "A fish gene in a strawberry? That can't be right." Old fears well up—of Chimeras, of monsters, of newborns with the wrong number of toes or an open spine, of the *bad seed* subtly undermining all that is normal and right. Once it is loose, you can't call it back. It is against nature. It is hubris. It will inevitably be punished.

(The fishy strawberry story happens to be wrong, but that doesn't affect its legendary impact. As usual with good stories, it's partly a conflation—in the early 1990s there was a *tomato* called Flavr-Savr with a bacterial gene to slow ripening, and it was popular, but it was too hard to ship after all and didn't taste great, so it disappeared; and there was an attempt in 1991 to duplicate a flounder gene in a tomato to make it frost tolerant, but it failed to work. There never was a strawberry with a fish gene in it.)

• The whole GE controversy was foreshadowed—and should have been settled—in the mid-1970s when recombinant DNA hit the fan. The technique, then called "gene splicing," emerged from a series of discoveries and inventions that culminated by 1974 in the ability to blend genes from various sources, including humans, into bacteria. One of the first applications was meant to be a cheaper, safer way to generate large quantities of insulin for treating diabetes. Everyone realized this was going to be an immensely powerful technology. Some of the scientists worried that potentially harmful strains of bacteria could be created, and they called for a moratorium on recombinant-DNA research until the hazards could be sorted out.

As soon as the issue got into the press, parts of the public began freaking out. Religious groups and environmentalists were up in arms. Scientists playing God, they declared, were committing abomination. Monstrous organisms would be created, environmentalists said, that could threaten everything living. There would be insulin-shock epidemics and tumor plagues. The Cambridge and Berkeley city councils—both cities the home of major universities—outlawed recombinant-DNA research. The U.S. Congress began introducing restrictive legislation.

That was the atmosphere that led to the Asilomar Conference on

Recombinant DNA Molecules in California in February 1975. Coming from all over the world, some 146 genetic scientists and related professionals convened for four days to regulate their research. They instituted an array of laboratory containment practices and mandated the use of organisms that could not live outside the lab. Some experiments were banned entirely, such as tinkering with the genes of pathogenic organisms. The guidelines were soon adopted and enforced in the United States by the National Institutes of Health.

Was Asilomar a good idea? The question was controversial then and remains controversial now. As it happened, I had an early window on the issue from an inside perspective.

One of my jobs as an assistant to California governor Jerry Brown was to arrange visits to his office by leading intellectuals. Brown's view was that "government may not always be the first to know about important new ideas, but it should not be the last." Thus every few weeks I got to spend a day hosting the likes of organizational guru Peter Drucker, futurist Herman Kahn, farmer-poet Wendell Berry, and media celebrator Marshall McLuhan. In 1977, two years after Asilomar, the California legislature was threatening to regulate recombinant DNA research in the state, so James Watson, the codiscoverer of the structure of DNA and director of the renowned Cold Spring Harbor Laboratory, came to visit. Watson had been an early supporter of the moratorium on recombinant DNA research and had helped to organize Asilomar.

In a short talk to a group including Brown, the governor's staff, and some legislators and press, Watson said:

> My position is that I don't regard recombinant DNA as a major or plausible public health hazard, and so I don't think that legislation is necessary. My experience is partly as a scientist who has worked with DNA for thirty years, and as a director of a laboratory where I have moral and legal responsibility for our work with viruses, and I think about it. . . .
>
> The general rule that seems to govern most of the people working with medical microbiology is that once you've domesticated a bacteria or a virus and got it to grow under conditions

of a laboratory culture, it loses its virulence. Biochemists generally live pretty long. (That's not true of organic chemists—they normally have a life span about five years shorter than other people, because those fumes that smell bad are bad.) . . .

We've been told there's a danger of *E. coli* getting a gene for cellulase via recombinant DNA. If there were a real selective advantage for *E. coli* to have cellulase, which would do away with our species [through starvation], it would have it already—by viral transfer or other means. Evolution is constantly trying to seek out the best set of genes for a given organism to occupy a niche. I do not worry about "monsters." . . .

Some people have said the [Asilomar] guidelines are capricious. I think they're totally capricious and totally unnecessary. We must have wasted $25 million on those precautions by now and it's on its way to $100 million. I think it's the biggest waste of federal money since we built all those fallout shelters. . . . It's silly to control where there's no evidence of danger. I am totally agreed that the public should participate in any process where they can be given facts to think about. But the tradition is, you don't call fire until you see it.

Watson was right, it turned out. The authoritative book on the history of molecular biology is Horace Judson's *The Eighth Day of Creation* (1996). A year after the Asilomar conference, Judson reports, "scientists' fears were receding fast. All the biologists who had signed the original appeal for the moratorium and the conference now agreed that the dangers had been exaggerated and the practical effects pernicious. Over two decades, the guidelines have been progressively relaxed." Recombinant-DNA techniques—genetic engineering—went on to revolutionize human medicine, transform every branch of biology, and become a major tool of disciplines ranging from chemistry to crime detection, from anthropology to agriculture. All without a single instance of harm.

So, was Asilomar a good idea? I would say yes, but for political reasons rather than scientific. The guidelines set by the scientists were far more specific and appropriate than politicians would have set, and those guidelines could be adjusted annually in response to real experience in the world, whereas political regulations not only resist fine-tuning, they defy any change at all. The recent simplistic legislation banning most human stem-cell research in the United States was a classic case. The Asilomar scientists forestalled that kind of folly by taking public responsibility themselves, early and adaptively.

● One particularly ingenious early adopter of the new genetic technology was Bruce Ames, a biochemist at the University of California–Berkeley. The problem he wanted to solve concerned the tens of thousands of novel chemicals that industry routinely creates and releases into the environment without much testing for their toxicity. An animal test for carcinogenicity using rats or mice would take two years and cost up to $750,000 *per chemical*, and a thousand new chemicals were being introduced every year.

Proposing that any chemical that would cause cancer would also cause genetic mutations, Ames engineered a strain of bacteria so that it would provide a quick test for mutagenicity as a stand-in for the carcinogenicity test. He took *Salmonella* bacteria and elegantly tailored them so they would starve in a petri dish filled with a particular nutrient unless they mutated the ability to eat it. He would drop a sample of the chemical to be tested in the dish. If, two days later, colonies of *Salmonella* were prospering in the dish, it showed that the chemical caused mutations and probably caused cancer. Though it took Ames ten years to develop the test, he gave away his *Salmonella* strain and his technique free of charge or patent to any who wanted it. Instead of the two years and hundreds of thousands of dollars required for an animal test, the Ames test took two days and cost $250 to $1,000. It became the world standard test for carcinogenicity and continues in use today.

The success of the Ames test was part of what inspired a major conflict inside Friends of the Earth, the environmental organization founded by Dave Brower in 1969. In the 1970s, while it was growing rapidly in size, impact, and international reach, Friends of the Earth launched a fight against genetic engineering. Two prominent members of the FOE advisory council argued against the campaign and then resigned in protest when their advice was ignored. They were Lewis Thomas, the celebrated author of *The Lives of a Cell* and head of the Sloan-Kettering Cancer Center, and Paul Ehrlich, then the most influential of Green-activist scientists. I happened to get a copy of Ehrlich's protest letter and printed it in *CoEvolution Quarterly*.

What Paul Ehrlich wrote in 1977 still applies:

Dear Friends:

As a professional biologist, I have become increasingly concerned about the opposition to recombinant DNA research expressed by FOE and some other environmental groups. . . . As an evolutionist, I have been rather depressed by various features of the ensuing controversy. It has been stated that the research should be discontinued because it involves "meddling with evolution." *Homo sapiens* has been meddling with evolution in many ways and for a long time. We started in a big way when we domesticated plants and animals. We continue every time we alter the environment. In general, recombinant DNA research does not seem to represent a significant increase in the risks associated with such meddling—although it may significantly increase the rate at which we meddle. . . .

One must always remember that any laboratory creations would have to compete in nature with the highly specialized products of billions of years of evolution—and one would expect the products of evolution to have a considerable advantage. In addition, there is evidence that bacterial species have been swapping DNA among themselves for a very long time and perhaps even exchanging with eucaryotic (higher) organisms.

In my view, the brightest promise of recombinant DNA is as a general research tool for molecular geneticists—a tool to use in the pursuit of the highest goal of science, understanding of the universe and of ourselves. . . . If recombinant DNA research is ended because it could be used for evil instead of good, then all of science will stand similarly indicted, and basic research may have to cease. If it makes that decision, humanity will have to be prepared to forego the benefits of science, a cost that would be high indeed in an overpopulated world utterly dependent on sophisticated technology for any real hope of transitioning to a "sustainable society." . . .

Of special interest to environmentalists and the FOE board is the work of Bruce Ames at Berkeley in developing tester strains of *Salmonella*, which are used in screening the products of the prolific synthetic organic chemical industry for carcinogens. Those of you who are familiar with it undoubtedly recognize that the "Ames screen" is the single most powerful technical tool that has been handed to environmentalists in decades. And yet Ames' research is being hindered by a complete—and pointless—ban on all recombinant DNA research with *Salmonella*. FOE is thus in the position of helping to block work on the most promising techniques for detecting environmental carcinogens.

◆

That was the moment when environmentalists turned away from reason in regard to genetic engineering. Thirty years after Paul Ehrlich resigned in protest, Friends of the Earth and all the other environmental organizations I know of still oppose genetic engineering. Most of all they oppose transgenic food crops; thus the great coinage "Frankenfood." Since pickiness about diet and loathing of the "wrong" food is such an ancient cultural practice, maybe that's the heart of the matter. I suspect that if environmentalists felt OK about eating genetically engineered food, their other complaints would fade away, so let's start there.

The most massive dietary experiment in history has taken place since 1996. One enormous set of people—everyone in North America—bravely ate vast quantities of genetically engineered food crops. (Some 70 percent of processed foods in the United States now have GE ingredients, mostly from corn, canola, and soybeans; nearly half of the field corn that goes into our corn muffins, corn chips, and tortillas is GE; and half of our sugar comes from GE sugar beets.) Meanwhile, the control group—everyone in Europe—made the considerable economic sacrifice of doing without GE agriculture and went to the further trouble of banning all GE food imports. It was great civilization-scale science, and the result is now in, a conclusive existence proof. No difference can be detected between the test and the control group.

Peter Raven sums up the outcome: "There is no science to back up the reasons for concern about foods from GM plants at all. Hundreds of millions of people have eaten GM foods, and no one has ever gotten sick. Virtually all beers and cheeses are made with the assistance of GM microorganisms, and nobody gives a damn."

(Who's Peter Raven? The head of the Missouri Botanical Garden for the last four decades, he is usually described as "one of the world's leading botanists and advocates of conservation and biodiversity." He has won nearly every environmental prize there is, as well as the National Medal of Science; *Time* declared him a "Hero for the Planet.")

Raven is hardly alone in his conclusions. In 2004 the United Nations Food and Agriculture Organization reported:

> The question of the safety of genetically modified foods has been reviewed by the International Council of Science (ICSU), which based its opinion on 50 authoritative independent scientific assessments from around the world. Currently available genetically modified crops—and foods derived from them— have been judged safe to eat, and the methods used to test them have been deemed appropriate.
>
> Millions of people worldwide have consumed foods derived from genetically modified plants (mainly maize, soybean, and oilseed rape) and to date no adverse effects have been observed. . . .
>
> Allergens and toxins occur in some traditional foods and can adversely affect some people leading to concerns that genetically modified plant-derived foods may contain elevated levels of allergens and toxins. Extensive testing of genetically

modified food currently on the market has not confirmed these concerns.

Conclusion: "To date, no verifiable untoward toxic or nutritionally deleterious effects resulting from the consumption of foods derived from genetically modified foods have been discovered anywhere in the world."

Here's one more overview, from a 2007 article titled "The Real GM Food Scandal," in Britain's *Prospect* magazine:

> The fact is that there is not a shred of any evidence of risk to human health from GM crops. Every academy of science, representing the views of the world's leading experts—the Indian, Chinese, Mexican, Brazilian, French and American academies as well as the Royal Society, which has published four separate reports on the issue—has confirmed this. Independent inquiries have found that the risk from GM crops is no greater than that from conventionally grown crops that do not have to undergo such testing. In 2001, the research directorate of the EU commission released a summary of 81 scientific studies financed by the EU itself—not by private industry—conducted over a 15-year period, to determine whether GM products were unsafe or insufficiently tested: none found evidence of harm to humans or to the environment.

• The science is in. The question then is, what do anti-GE environmentalists do with it? Some just deny the science exists. In February 2008 the editor of a UK magazine inaccurately named *The Ecologist* wrote, "[Politicians] can truthfully say that they have never seen any data showing that eating GM is harmful to humans, because, of course, the research has never been done." Others say the science is irrelevant. Joschka Fischer, a leader of the Green Party and Germany's foreign minister from 1998 to 2005, said in 2001, "Europeans do not want genetically modified food—period. It does not matter what research shows; they just do not want it and that has to be respected."

The weird fact is that GE foods, because they are so exhaustively vetted, are safer to eat than the products of conventional and organic farming. In the United States, new GE crops are tested by the Food and Drug Administration, the National Institutes of Health, *and* the Environmental Protection Agency, whereas new crop varieties created by breeding go through no such process.

Consider the kiwi. Arrived at by traditional selective breeding of the Chinese gooseberry and marketed aggressively by New Zealanders, it has become popular worldwide. Yet, as Israeli plant scientist Jonathan Gressel points out, "If the kiwi fruit had been genetically engineered, it would not be on our tables. A very small proportion of the population develops severe allergies to kiwi fruit with a wide range of symptoms, from localized oral allergy syndrome to life-threatening anaphylaxis, which can occur within minutes after eating the fruit." (Peanuts, shellfish, wheat, dairy products, and other common foods also cause allergies; they—and the kiwi—could some day be made nonallergenic with genetic engineering, and should be. A nonallergenic GE peanut is already being developed at the University of Georgia.)

Toxicity is normal in plants because they can't run from their predators and competitors. They have to stand and fight. They use spines against the predators, shade against the competition, and poison against everybody, including us. Organic farmer Raoul Adamchak reports, "I have discovered that green potatoes make pretty good rodent poison. One day I went into the certified organic hoophouse to find three dead mice near some freshly eaten green potatoes." His wife, geneticist Pamela Ronald, comments, "So far, compounds that are toxic to animals have only cropped up in foods developed through conventional breeding approaches. There have not been any adverse health or environmental effects resulting from commercialized GE crops."

Bruce Ames once did animal carcinogenicity tests on 27 of the 1,000-plus natural chemicals in a sample of roasted organic coffee. He and his colleague Lois Gold found that 8 of the chemicals were not harmful, but 19 caused cancer in the rats—acetaldehyde, benzaldehyde, benzene, benzofuran, and on through styrene, toluene, and xylene. "On average," they wrote, "Americans ingest roughly 5,000 to 10,000 different natural pesticides and their breakdown products."

Geneticist Nina Federoff elaborated in *Mendel in the Kitchen* (2004):

> Lima beans contain a chemical that breaks down during digestion into hydrogen cyanide, which is poisonous. Toxic psoralens in celery cause skin rashes. Moreover, psoralen cross-links the strands of DNA to each other, which can cause cancer. A chemical in cauliflower can make the thyroid enlarge. Carrots contain a nerve poison and a hallucinogen. Peaches and pears promote goiters. Strawberries contain a chemical that prevents blood from clotting and can lead to uncontrollable bleeding. Peas, beans, cereals, and potatoes contain lectins, which cause nausea, vomiting, and diarrhea.

Everything depends on dosage. It takes four hundred carrots to give you a harmful dose of neurotoxin. (I know of a food zealot who single-handed his sailboat from California to Hawaii eating only carrots along the way. By the time he arrived, his eyes were orange and he was hallucinating.)

Nevertheless, eat your vegetables. Along with all that mild poison are crucial micronutrients and antioxidants. Bruce Ames found that one of the major causes of cancer is dietary imbalance: "The quarter of the population eating the least fruits and vegetables has double the cancer rate for most types of cancer compared with the quarter eating the most." (Some hormesis theorists speculate that vegetables are good for us *because* of the very-low-dose natural toxins keeping our detoxifying mechanisms busy and fit.)

It's worth remembering that in Mary Shelley's story, Dr. Frankenstein's creature was a misunderstood good guy, wrongly vilified by the fearful populace. Of course, that's a rhetorical argument, devoid of meaning. But so is the term *Frankenfood*.

✦

One reason biologists are unalarmed about genetic engineering is that they know what a minor event it is amid the standard chaos of evolution and the just-barely-organized chaos of agricultural breeding. Take our own genetic makeup, for instance. Science journalist Carl Zimmer notes: "Scientists have identified more than 98,000 viruses in the human genome, along with the mutant vestiges of 150,000 others. . . . If we were to strip out all our transgenic DNA, we would become extinct." (More on that matter in the next chapter.)

Human evolution has been accelerating ever since we left Africa seventy thousand years ago, and it sped up even further once we began coevolving with agriculture. Races emerged only twenty thousand years ago. With the larger population that agriculture permitted, our rate of evolution increased a hundredfold. Seven percent of our working genes are recent adaptations. "Nobody 10,000 years ago had blue eyes," says anthropologist John Hawks. "Why is it that blue-eyed people had a 5 percent advantage in reproducing compared to non-blue-eyed people? I have no idea." (My blue-eyed hypothesis would be that a number of males and females decided that blue eyes were sexy.) Gregory Cochran, an evolutionary biologist at the University of Utah, says that "history looks more and more like a science fiction novel in which mutants repeatedly arose and displaced normal humans—sometimes quietly, by surviving starvation and disease better, sometimes as a conquering horde. And we are those mutants." Thanks to globalization and urbanization, races everywhere are mixing more, and that gives evolution even more variability to work with. We are becoming a world of smart mutts.

Craig Venter, the leading cataloger of the human genome, notes that only 3 percent of our genome consists of protein-producing genes, and the rest is "regulatory regions, DNA fossils, the rusting hulks of old genes, repetitious sequences, parasitic DNA, viruses, and mysterious stretches of who-knows-what." (Venter is careful never to use the term *junk DNA*.) One truly selfish gene has made a million copies of itself, taking up 10 percent of our genome; it seems to have no function other than self-replication. Genetically speaking, humans are a fast-moving mess. So is everything else.

● Humanity's first venture into genetic modification—agriculture—was a global event. We exploited the genetic malleability of dozens of plants in at least ten independent centers of agricultural innovation. The process was gradual, progressing from selective gathering to small-scale and then large-scale clearing and tilling, to techniques such as irrigation and crop rotation for what were by then exquisitely designed cultivars. Sing their praises!—the crops of the Americas: squash (first domesticated 10,000 years ago), corn (9,000 years), potatoes (7,000 years), peanuts (8,500 years), and chilis (6,000 years); the crops of the Mideast: rye (13,000 years), figs (11,400 years), wheat (10,500 years), and barley (10,000 years); the crops of China: rice (8,000 years) and millet (8,000 years); the crops of New Guinea: bananas (7,000 years), yams (7,000 years), and taro (7,000 years); and the crops of Africa: sorghum (4,000 years) and pearl millet (3,000 years). Humans rearranged thousands of plant genes, and the world was permanently transformed.

Some anti-GE activists talk of defending the "intrinsic integrity" of crop-plant genomes. What integrity? Crop plants have no integrity of their own; they are products of human tinkering and only remotely resemble their wild cousins. Botanist Klaus Ammann points out that good old wheat, fashioned through good old breeding, has modifications that include "the addition of chromosome fragments, the integration of entire foreign genomes, and radiation-induced mutations."

Transgenic blending is an old story in agriculture. As philosopher Johann Klassen argues, "We don't feel revulsion at the thought of a mule, an unnatural cross between horse and donkey; at a rutabaga, an unnatural cross between cabbage and turnip; at triticale, an unnatural cross between wheat and rye."

Organic farmer José Baer (son of solar pioneer Steve Baer) offers perspective:

> We have been creating weird crop species for millennia. First we simply selected the ones we liked and planted them again, then we used chemical mutagenesis to speed up mutations and arrive at the ones that weren't likely to appear anytime soon. Then we discovered radiation mutagenesis was even more effective. GM allows us to pick the trait and splice it in

rather than discovering the trait we were looking for in amongst the rubble of mutated seedlings.

As a certified organic farmer, Baer grows several crop varieties that come from radiation mutagenesis—blasting seeds with intense radiation to get more mutations in the seedlings—widely used by breeders since 1927 and common on organic farms, but he is not allowed to grow any genetically engineered crops. A comparison study published in 2007 by the National Academy of Sciences showed that rice lines resulting from gamma radiation mutagenesis caused more disruption of "non-target genes"—more unexpected consequences—than occurred in transgenic lines of rice. Organic rules force José Baer to use seed with a slightly greater risk of unintended consequences.

All forms of breeding introduce unexpected consequences, whether or not they're accelerated by chemicals or radiation (as two thousand current crop varieties have been). "A standard problem with breeding," says plant geneticist Pamela Ronald, "is you pull a trait from one plant over to another, and you get linked traits that you don't want. It's called 'linkage drag.' With genetic engineering you don't get that." The precision of GE, writes Jonathan Gressel, lets you control "in which tissue, under what circumstances, and how much a gene will be expressed. This is done without introducing all the extraneous genetic baggage brought by crossing with the related species. Genetic engineering is like getting a spouse without in-laws, whereas breeding is like getting a spouse with a whole village." The traditional method for removing the extraneous genes is with many generations of backcrossing—a blind, random, painstaking process.

Unwelcome traits still get through. In the 1960s the Lenape potato was developed by crossing the popular Delta Gold potato with a wild relative to increase insect resistance. The Lenape was delicious and indeed insect resistant, so it was released to public use, distributed as popular potato chips and used by other breeders. But after one breeder found that Lenapes made him nauseous, analysis showed that they were high in a natural glycoalkaloid toxin from the wild potato. The Lenape was formally withdrawn, but it was too late. Thirteen varieties of potatoes still remain on the market with Lenape toxins bred into them.

● In 1999 Amory Lovins made a public statement against GE that puzzled me. (I'm using Lovins as a foil more than he deserves because his statements have such clarity that they offer good purchase for argument.) "Shotgunning alien genes into the genome," he wrote, "is like introducing exotic species into an ecosystem." If Lovins's background was in biology instead of physics, I doubt he would have written that. His analogy between a gene and a whole species is misplaced. Genes may be selfish, but they are nowhere near as freestanding and robust as wild species are. A gene is a fragment, one among the tens of thousands of genes it takes to make a whole organism. No individual gene can dominate or transform a genome the way some alien-invasive species can dominate an ecological neighborhood. A gene has to go along with most of the game in order to function at all. A more useful analogy would be to a computer program that has a bug. GE is a bug fix—a tiny bit of the right code in the right place to make a problem go away.

I think Peter Raven nailed what it's really all about. "When people talk about taking genes from one distantly related organism to another," he told an interviewer in 1999, "they talk as if every gene in a mouse had a little mouse in it and putting it somewhere else would be bizarre." Genes, he said, are nothing but "strings of bases which, in triplets, specify the amino acids that make up proteins. A lot of different organisms use similar or nearly identical genes to do the same job."

Where a gene comes from is irrelevant; the point is what it does.

Genetic engineering is so much more precise, transparent, and accountable than breeding, it invites the thought experiment proposed by technology historian Kevin Kelly: "Suppose the sequence was reversed. Suppose genetic engineering is what we had been doing all along. Then some group says, 'No, we're going to use this new process called *breeding.* We'll create all kinds of interesting recombinations, we'll blast seeds with radiation and chemicals to get lots of mutations, and we'll grow whatever comes up, pick the ones we like, and hope for the best.' What would people say about the risk of doing it that way?" They would call it genetic gambling, says inventor Danny Hillis, and outlaw it.

As for medicine, the panic generated by recombinant DNA back in the 1970s has completely died away. In 1982, a human gene was introduced into the bacterium *E. coli* so that "bioreactors" of trillions of the engineered organisms could generate vast quantities of human insulin in a way that is safer and far cheaper than the old technique using the pancreases of calves and pigs. These days about a quarter of all new drugs are made by genetic engineering—130 so far in the United States, 87 in Europe—and the rate is accelerating. We put these substances into our bodies without a second thought, and for good reason. GE drugs are just as safe as GE foods.

Actually, the foods are safer, free of the overdose and bad-combination risks that all drugs have.

✦

"Natural food." Whenever I hear those words crooned, my inner crank cranks up. "Natural!" I would rail if I gave it voice. "No product of agriculture is the slightest bit natural to an ecologist! You take a nice complex ecosystem, chop it into rectangles, clear it to the ground, and hammer it into perpetual early succession! You bust its sod, flatten it flat, and drench it with vast quantities of constant water! Then you populate it with uniform monocrops of profoundly damaged plants incapable of living on their own! Every food plant is a pathetic narrow specialist in one skill, inbred for thousands of years to a state of genetic idiocy! Those plants are so fragile, they had to domesticate humans just to take endless care of them!"

To an ecologist, or to a Gaian for that matter, agriculture is one vast catastrophe. The less of it, the better. Thus Peter Raven: "Nothing has driven more species to extinction or caused more instability in the world's ecological systems than the development of an agriculture sufficient to feed 6.3 billion people." Thus Jim Lovelock: "The fact that at least 40 percent of the land surface is used for food crops is hardly ever taken into account in our current approach to climate change. A self-regulating planet needs its ecosystems to stay in homeostasis. We cannot have both our crops and a steady comfortable climate." (That 40 percent of the Earth's land breaks down to 5.8 million square miles in permanent cropland, 13.5 million square miles in permanent pasture. The remaining 31 million square miles is ecological.)

Opponents of genetic engineering are right to suspect GE crops of being ecologically harmful, because all crops are ecologically harmful. The question then becomes: How do GE crops compare with traditional crops in terms of doing ecological damage or ecological good? By "good" I mean richer soil, more wildlands, better integration with noncrop biodiversity, and (adding a social goal) getting more people the hell out of poverty and away from starvation. Critics of GE focus on soil effects, on quantities of herbicides and pesticides in the environment, and on the potential for creating superweeds and superbugs. Exhaustive studies have been made on all these questions, and the data is in.

● About 40 percent of crop yield in the world is lost to weeds and pests every year. The spectacular success of GE crops in lowering those losses is why *Science* magazine could report in 2007, "Over the past 11 years, biotech crop area has increased more than 60-fold, making GM crops one of the most quickly adopted farming technologies in modern history." Just two GE traits are responsible for most of that success—herbicide tolerance and insect resistance.

The herbicide in question is glyphosate (pronounced GLY-fo-sate), discovered in 1971 and marketed since 1974 by Monsanto as Roundup. It's a pretty miraculous compound. Sprayed on a plant's leaves, glyphosate disables an enzyme in the chloroplasts so that the plant starves to death over a week or two. It has no proven effect on any animal—insects, fish, birds, mammals, or us. It binds in the soil and degrades to harmlessness within weeks, so it doesn't pollute the water or linger in the ground like other herbicides, and it has a fraction of their toxicity. Starting in 1996, Monsanto introduced "Roundup Ready" corn, soybeans, cotton, canola, sugar beets, and alfalfa, all genetically engineered to tolerate glyphosate. Then the price of the herbicide dropped by half when glyphosate went out of patent in 2000. The attractions of cheap glyphosate overwhelmed traditional farmer conservatism. By 2007 in the United States, over 90 percent of soybeans and 75 percent of corn crops were engineered for glyphosate tolerance.

The great ecological win with GE herbicide-tolerant crops is that they encourage what is called no-till agriculture. Farmers don't have to plow. The stubble from last year's crop rots in the field, turning into compost and habitat for small wildlife, and the soil is held in place instead of eroding away. Farmers employ a method called direct seeding to inject seeds, each with a shot of fertilizer, into the soil through the stubble; then, just when their crops begin to emerge, they spray the field with glyphosate and wipe out all the weeds with no effect on the glyphosate-tolerant crop plants. The result is a high-yield crop and intact soil that grows in richness from year to year, full of life, with a lovely crumb structure. "With no-till," says Jim Cook, a plant pathologist and sustainable-agriculture evangelist at Washington State University, "you improve soil structures, stop erosion, sequester carbon, improve water filtration rather than letting it run off the land, and store more water in years of drought."

There are also major climate benefits. Soil holds more carbon in it than all living vegetation and the atmosphere put together. (Earth's soil holds about 1,500 gigatons of carbon, versus 600 gigatons in living plants and 830 gigatons in the atmosphere.) Tilling releases that carbon. Jim Cook explains: "Carbon disappears faster if you stir the soil. If you chop the crop residue up, bury it, and stir it—which is what we call tillage—there's a burst of biological activity, since you keep making new surface area to be attacked by the decomposers. You're not sequestering carbon anymore, you're basically burning up the whole season's residue."

Plowed land is the source of gigatons of carbon dioxide in the atmosphere. Cultivated soil loses half of its organic carbon over decades of plowing, but sustained no-till farming can bring the carbon content back to a level the equal of wildland soil, such as in tallgrass prairies, according to soil microbiologist Charles Rice, who has done extensive comparative studies at Kansas State University. More and more of GE agriculture is shifting to no-till—80 percent of soybean acreage, for example—because it saves the farmer so much time, money, and fuel. Saving the fuel also helps save the atmosphere. "About 5 percent of all fossil fuel use is by agriculture, and most of this goes on weed control," writes geneticist Jennifer Thomson in *Seeds for the Future*.

Organic farmer Raoul Adamchak told me that the GE farmers he met at a no-till meeting in California are starting to sound like organic enthusiasts, bragging about the richness of their soil and counting their earthworms. Organic farmers, however, can't use GE, so they continue to plow every spring, burying the cover crops and early weeds, releasing the soil's carbon dioxide. (Some organic farmers are aware of the problem but haven't yet found a non-GE solution.)

● One very proper subject for criticism of any new kind of crop is how well or badly it fits in with the local ecology—the agroecology, as we say now. This is where the subject of "superweeds" comes up. It's a familiar aggravation for farmers. Weeds adapt quickly to exploit the sunlit, watered, fertilized luxuries of cropland and find it easy to compete with the coddled crop species if they can just dodge around the farmers' countermeasures. Some weeds, if they're related to the crop plants, may borrow helpful genes from them. That has happened historically with rice, millet, sorghum, sunflowers, and sugar beets, and it is legitimately something to worry about with new GE crop genes.

New glyphosate-resistant weeds *are* turning up, not so much through gene borrowing as through the usual evolutionary response to the increased selection pressure of any highly successful, overemployed, deadly technique. When only glyphosate is used to kill weeds, some weeds will evolve a workaround all too soon. The main countermeasure is to employ the full arsenal of integrated pest management—crop timing, crop rotation, biocontrols, etc.—so the weeds face too many simultaneous obstacles to evolve around. Genetic engineering also can "stack" genes for resistance to other effective herbicides such as Dicamba along with the glyphosate resistance, and the genes can be parked in the maternal portion of the genome so they can't spread to the world through the male pollen. Such seeds are in the pipeline. GE is just another tool in the ever-evolving integrated pest management kit.

Agricultural weeds specialize in threatening farmers, not the wildlands. A glyphosate-resistant weed in the anything-goes combat of the woods is like a boxer in a gunfight—wrong skills for the situation.

• What about superbugs? Here the questions are raised around the hugely popular engineered crops Bt corn and Bt cotton. The Bt stands for *Bacillus thuringiensis*, a common soil bacterium with a unique toxin lethal to the larvae of the butterflies and moths that normally feed on the now-GE crops. Organic farmers have long sprayed dried Bt bacteria on their crops because they are considered a natural pesticide; they are very effective; and they kill only the target insects, with no harm to beneficial insects or human customers. When the bacterial gene that makes the Bt toxin was engineered into corn and cotton, the transgenic plants were able to kill the pests by themselves: no spray necessary. Furthermore, because the toxin is in the plant, even spray-eluding corn borers are defeated. (That's why ears of organic corn usually have a live corn earworm in them. Bt corn is forbidden to organic farmers, so most sell no corn at all.)

The main ecological effect of Bt crops has been a drastic reduction in pesticide use. Cotton is especially pesticide-intensive. The introduction of GE cotton has dropped pesticide use by half. In one GE crop region studied, a tenth as many farmworkers are being hospitalized for ailments caused by farm chemicals. And, astonishingly, no insect pests significantly resistant to Bt crops have evolved yet, even though some five hundred insect species have developed resistance to sprayed pesticides.

Thus GE crops help mitigate greenhouse gases and are more ecologically benign than non-GE crops. "GM corn, cotton and soybean have been in commercial use for over five years now, and millions of hectares have been grown without any field reports of adverse ecological impacts," said a 2001 study in the peer-reviewed journal *EMBO Reports*. "Substantial environmental benefits," it continued, "have been established for some of these products, such as Bt cotton, because of the resulting reduction in the use of chemical insecticides. . . . Indeed, populations of predatory arthropods that help to control secondary pests like aphids are found to be consistently higher in Bt cotton fields than in sprayed fields of conventional cotton."

Peter Raven sums it up: "To assert that GM techniques are a threat to biodiversity is to state the exact opposite of the truth. They and other methods and techniques must be used, and used aggressively, to help build sustainable and productive, low-input agricultural systems . . . around the world."

✦

Many environmentalists argue that genetically engineered crops threaten the "family farm." The argument requires maintaining two illusions—one nostalgic, the other economic and political, both illusory because of historical forces that have nothing to do with GE.

The family farm idea is a bucolic fantasy based on selective memory, songs from musicals ("Howdy, neighbor! Happy harvest!"), and the marketing images used for organic produce. It recalls what the world was like in 1900, when half of all Americans lived and worked on farms. One percent of Americans now work on farms. You can't have a family farm if all the children light out for the city, and that's what they do. The average age of American farmers is fifty-five.

Government has tried to keep people down on the farm with an elaborate system of subsidies—some of them, perversely, supporting bad farming. Since rural values are by and large conservative values, you get the farce of conservative politicians legislating huge government subsidies and conservative farmers sitting in dying towns grumbling over coffee about the dole they're on.

The reality of farmwork is that it is the lowest-paid labor there is, and most of it is backbreaking. In the developing world, the toil is done largely by unpaid family members, mostly women, hacking away with hoes at the weeds in the tomato fields all day every day, or bent over in the rice paddies hand-planting seedlings in the muck. In the developed world we hire immigrant laborers to do it.

Environmentalists keep insisting that GE crops are bad for farmers, especially small-scale farmers. That position is tough to maintain as the reality emerges that even in the face of opposition, GE crops are tremendously popular. "Every time a GE crop has been approved for use," notes

Pamela Ronald, "farmers have embraced it and the GE acreage for each crop has quickly grown to 50 to 90 percent of the total acreage." A 2008 study funded by the Rockefeller Foundation reported:

> Two million more farmers planted biotech crops last year, to total 12 million farmers globally enjoying the advantages from the improved technology. Notably, 9 out of 10, or 11 million of the benefiting farmers, were resource-poor farmers. . . . In fact, the number of developing countries (12) planting biotech crops surpassed the number of industrialized countries (11), and the growth rate in the developing world was three times that of industrialized nations (21 percent compared to 6 percent).

In the developing world, nearly all farms are smaller than five acres, and most are one acre or less.

● Farmers want GE technology for their crops; nonfarmers want them not to want it. I've seen a classic case of that standoff play out between Marin and Sonoma counties in California. Our tugboat home is in nonagricultural Marin, close to San Francisco and its attitudes. Marin outlawed GE agriculture in 2002. The defunct dairy farm where we spend weekends is in Sonoma, the next county to the north, still primarily agricultural. In 2005 Sonoma voted on an ordinance to abate all GE agriculture in the county. The official argument for the measure evoked the standard language: "Measure M will protect Sonoma County's family farmers, gardeners and environment from irreversible genetic contamination by GE organisms. Our children should not be used as guinea pigs for genetic engineering. When GE crops are released, they create herbicide-resistant superweeds and contaminate the local food supply and natural environment. Genetic contamination is forever—it can't be recalled, contained or cleaned up."

Nine local farmers' organizations, including the Sonoma Country Farm Bureau, opposed the ban. "In contrast," Pamela Ronald notes, "urban residents, food processing companies, and wineries support[ed] it, hopeful to include 'GE-free Sonoma' on their label, as a new way to market their products." They lost: Measure M was defeated, 55 percent to 45 percent. Wherever farmers rule, GE wins. All the major agricultural counties in the state defeated similar resolutions.

In 2006, when two hundred French anti-GE activists destroyed fifteen acres of GE corn near Toulouse, eight hundred local farmers marched in a nearby town to protest the attack and petition the government to support GE research. In 2000, GE soybeans were legal in Argentina but outlawed in Brazil. The difference in productivity was so obvious that Brazilian farmers smuggled the seeds across the border, until their government relented and legalized GE agriculture.

● Why do environmentalists want to deny the advantages of GE crops to farmers in the developing world, who need it most? Robert Paarlberg, author of *Starved for Science: How Biotechnology Is Being Kept Out of Africa* (2008), theorizes that rich countries have the luxury of debating the nuances of economics and perceived risk around GE crops, whereas poor countries don't:

> The technology is directly beneficial to only a tiny number of citizens in rich countries—soybean farmers, corn farmers, a few seed companies, patent holders. Consumers don't get a direct benefit at all, so it doesn't cost them anything to drive it off the market with regulations. The problem comes when the regulatory systems created in rich countries are then exported to regions like Africa, where two thirds of the people are farmers, and where they would be the direct beneficiaries.

Florence Wambugu puts it baldly: "You people in the developed world are certainly free to debate the merits of genetically modified foods, but can we please eat first?" (A Kenyan plant pathologist educated on three continents, Dr. Wambugu did a three-year postdoc in genetic engineering with Monsanto, developing a virus-resistant GE sweet potato. She heads the Africa Harvest Biotech Foundation, based in Nairobi. In 2008 she won the Yara Prize for a Green Revolution in Africa.)

Poor farmers in the developing world take to GE seeds as readily as they take to cellphones, as a direct route out of poverty. The combination allows them to sell food to the new urban populations at urban prices and escape the cashless trap of subsistence farming. A similar flip happens at the national level. When, despite fierce lobbying by Indian pesticide companies, farmers in India adopted Bt cotton in 2002, the nation went from a cotton importer to an exporter, from a crop of 17 million bales to 27 million bales. What was the social cost of that? The 2008 Rockefeller-funded report says:

> A study of 9,300 Bt cotton and non-Bt cotton-growing households in India indicated that women and children in Bt cotton households have slightly more access to social benefits than non-Bt cotton growers. These include slight increases in pre-natal visits, assistance with at-home births, higher school enrollment for children and a higher proportion of children vaccinated.

The main event was that Bt cotton increased yield by 50 percent and diminished pesticide use by 50 percent, and the Indian growers' total income went from $840 million to $1.7 billion.

• Activists have been unable to persuade farmers that GE crops are bad for them because the farmers' direct experience tells them otherwise. So the activists focus instead on undermining the farmers' market by frightening their customers, the general public, with phantasms of genetic "contamination."

There are vivid stories that anti-GE people repeatedly tell each other—and anybody who will listen—to keep themselves reassured that their cause is just and their passion necessary. The stories become legend—the one about the monarch butterflies, the one about native Mexican corn, the one about the "terminator gene." On the other side, people who promote GE have their own tales: the papaya story, the "golden rice" story, the Zimbabwe poison story. They all have basis in fact, but in each case, important information is left out of the legend, especially the then-what-happened part. Pursued down to detail level, the stories shed a different light. Let's peruse.

BT CORN POLLEN KILLS MONARCH BUTTERFLIES

The story: A May 1999 issue of *Nature* included a one-page note— "Transgenic Pollen Harms Monarch Larvae"—by Cornell entomologist John Losey and colleagues. They reported a four-day lab experiment in which monarch caterpillars were fed an unspecified amount of pollen from Bt corn, and 44 percent of them died. The story made the front page of the *New York Times* and was broadcast widely by GE opponents. It still is.

The rest of the story: Subsequent exhaustive field research showed that the actual effect of Bt corn pollen would kill, at most, three monarch caterpillars out of ten thousand, a minute fraction of the hazards monarchs face from other impacts of civilization. Six detailed papers on the subject (with thirty authors) appeared in the *Proceedings of the National Academy of Sciences* (PNAS) in September 2001 but got no secondary press because their news was lost in the media din created by 9/11. Superenvironmentalist Peter Raven cowrote a July 2000 paper in *PNAS*, which concluded that, because the major real harm to monarch populations is from habitat loss and the use of pesticides in Mexico and the United States, and "considering the gains obviously achieved in the level of survival of populations of monarch butterflies and other insects by eliminating a large proportion of the pesticides applied to the same crops, the widespread cultivation of Bt corn may have huge benefits for monarch butterfly survival."

(Monarchs are bright orange for an interesting coevolutionary reason. The larvae feed on milkweeds, having developed the talent to endure the poison, a cardiac glycoside, with which the plants repel most other bugs. The monarch caterpillars not only tolerate the poison, they incorporate it into their brightly colored bodies and pass it on to their brightly colored, slow-flying adult form. The brilliant colors are advertising: "I'm really poisonous!" Birds learn to shun monarchs. Another somewhat poisonous butterfly, the viceroy, which gets *its* poison from eating willow leaves, adapted to look exactly like the monarch so that birds would efficiently lump all vile-tasting prey into a single orange avoidance image. It's one of the classic just-so stories of coevolution. The concept of coevolution, by the way, was coinvented in 1964 by botanist Peter Raven and lepidopterist Paul Ehrlich—before his population bomb fame. Their paper, "Butterflies and Plants: A Study in Coevolution," is one of the most cited in biology. When Peter Raven says Bt corn is good for monarchs, he probably knows what he's talking about.)

GENETIC POLLUTION: BIOTECH CORN INVADES MEXICO

The story: In November 2001, *Nature* published "Transgenic DNA Introgressed into Traditional Maize Landraces in Oaxaca, Mexico," by David Quist and Ignacio Chapela. The paper's impact was immediate and widespread, because landraces are important to agriculture. They are reservoirs of genetic diversity, maintained by traditional, small-scale farmers who save and select their seeds for optimization to local tastes and local conditions. Particularly worrisome is anything suspicious happening with corn landraces in Mexico, where the crop was first created and first diversified; landraces comprise two thirds of all the corn grown in Mexico.

The *Nature* paper suggested that genes and gene fragments from GE corn were turning up in landrace genomes, uninvited and through untraceable pathways. "This is the world's worst case of contamination by genetically modified material because it happened in the place of origin of a major crop," said Jorge Soberón, secretary of Mexico's National Biodiversity Council. "It was as if someone had gone to the United Kingdom and started replacing the stained-glass windows in the cathedrals with plastic."

The rest of the story: The anti-GE world exploded with outrage about transgenic genes "polluting and degrading one of Mexico's major treasures" and the "Gene Giants . . . insulting the socio-cultural rights of Mexican farmers." Greenpeace called for a ban on GE corn in Mexico.

Then *Nature* published two letters criticizing the methodology of the original Quist and Chapela paper along with an editorial discrediting the paper and regretting its publication. Four years later, in August 2005, Jorge Soberón was one of six authors (four in Mexico, two in the United States) of a paper in *PNAS* titled "Absence of Detectable Transgenes in Local Landraces in Oaxaca, Mexico (2003–2004)." The authors, "convinced we were going to verify Quist and Chapela's results," examined 153,746 seeds from 125 fields in the same region and found nothing. Despite the ongoing presence of GE corn in Mexico, there were *no* GE genes in the landrace corn crops. The authors speculated that if GE genes had been present in 2000, they had diluted to undetectable levels by 2003. (Science marches on: A researcher in 2008 reported finding traces of transgenes in 1 percent of 2,000 landrace samples.)

The whole episode was good for crop science because it set in motion important research into what is called "gene flow," and corn is a fine subject for study because, unlike most crop plants, it is a promiscuous outcrosser. The maize landraces, it turns out, have been swapping genes with commercial crops for decades with no loss of diversity; GE genes are expected to be no different in that respect. Landrace farmers are well aware of the problem of "inbreeding depression" (as they say, the maize "gets tired"—*se cansa*), so they routinely blend in other varieties, and also vagrant genes on the pollen are always blowing from cornfield to cornfield. Israeli plant scientist Jonathan Gressel describes the customary situation in *Genetic Glass Ceilings: Transgenics for Crop Biodiversity* (2007):

> There has been gene flow from commercial varieties of crops to/from landraces growing nearby, only to the betterment, at times, of one party or the other. The farmer preserves the landrace, morphologically, tastewise, but actually (inadvertently) selects for individuals that have also picked up genes for disease or stress tolerance, or higher yields. This is especially apparent with the steadily improving maize landraces selected by Mexican farmers. The landraces of a century ago are not

genetically identical with those two centuries ago or today,
even if the farmers think they are.

Gressel deplores "politically laden terms such as preserving 'genetic purity,' which has the same connotation as preserving 'racial purity.' " The real threat to landrace diversity everywhere in the world is urbanization—the young leave for better jobs in the city, and the local crops die out. Then the one remaining hope for abandoned landrace genomes is seed repositories like the famed International Maize and Wheat Improvement Center, near Mexico City.

Gene flow is the norm, in agriculture and in nature. Transgenes *will* flow, causing no more harm than any other commercial crop genes, and no less. There are three common solutions for the gene-flow problem, one easy, two tricky. One tricky technique involves making sure the implicated genes can't flow because they're engineered out of the pollen. Another approach involves surrounding the GE crops with "refugia" of non-GE plants, which isolates the transgenes. The easy fix simply makes sure that the GE plants are sterile, incapable of reproducing. Zero gene flow: complete solution. GE opponents *hate* that one. Read on.

TERMINATOR GENE FROM MONSANTO ENSLAVES FARMERS

The story: In 1998 a patent was awarded to the Delta & Pine Land Company for a technique they called GURT—genetic use restriction technology—which would engineer sterility into GE crop plants. The idea was to prevent farmers from cheating on their legal agreements with seed suppliers to always buy new seeds rather than keeping some for replanting. The agritech giant Monsanto, which owned a piece of Delta & Pine Land, indicated that it might buy the whole company to obtain a better stake in cotton seed supply. This raised fears that Monsanto might apply the GURT sterility technique to its GE crops.

A cry went up worldwide from GE opponents. "The ultimate goal of genetic seed sterility is neither biosafety nor agronomic benefits, but bioserfdom," declared Pat Mooney, coining a term that stuck—*terminator technology.* (Mooney was then head of Rural Advancement Foundation International, later renamed ETC Group.) "This is an immoral technique that robs farming communities of their age-old right to save seed and their role as plant breeders," said a Chilean official. "This is the neutron bomb of agriculture." The roar grew for a year, especially in Europe and the developing world. Eventually Monsanto announced that it would not pursue the technology. (Sir Gordon Conway at the Rockefeller Foundation and Florence Wambugu in Kenya were among those who urged Monsanto to back off.) As of 2009, there were no sterilized GE crops anywhere in the world, handy as they would be for stopping unwelcome gene flow.

The rest of the story: Score one for impoverished environmentalists outmarketing rich corporations with great scare language like "terminator" and "suicide gene." (More accurate would have been "maiden aunt technology," but good luck selling that one.) There was some overreach though, such as this remarkable sentence: "The gradual spread of sterility in seeding plants would result in a global catastrophe that could eventually wipe out higher life forms, including humans, from the planet." When I read that line somewhere, quoting Vandana Shiva, the antiglobalization activist in India, I figured it must be a misquote, so I looked it up in her book, *Stolen Harvest: The Hijacking of the Global Food Supply* (2000). There it was; check it yourself. On page 83, you'll find her warning of humanity's doom from *heritable sterility*—a biological impossibility. Shiva usually describes herself as having been one of India's leading physicists.

(By the way, the line is not originally hers. It was borrowed verbatim and unattributed from a 1998 essay privately published by Geri Guidetti, who runs the Ark Institute, a survivalist seed supply service in Oregon. According to one online description, Guidetti "has taught the biological and biochemical sciences at universities across the US for 20 years. She has authored hundreds of science and research articles." Maybe so, but all I could find online by her were interviews and essays about the dangers of Y2K, bird flu, terrorists, and terminator technology.)

The fear that GE sterility technology would require the annual purchase of seeds is less novel and less alarming when viewed in the context of standard agricultural practice. Most farmers buy vigorous new hybrid seeds every year and have for decades. Hybrid seeds don't "breed true": the next generation is a chaotic mix. "From the seed company's point of view," explains Raoul Adamchak,

> this is great, for each year the hybrid seeds have to be created anew by the seed company. They are expensive, but most organic growers buy them, because the hybrid vigor, uniformity, disease resistance, yield, and sometimes taste, are deemed to be worth the extra cost. And, most farmers are unwilling to create their own inbred lines by cross-pollination each year. Few have the time to be both a breeder and a farmer.

Following the development of hybrid corn in the 1920s, by 1970 some 96 percent of U.S. corn crops were hybrid, and yield went up from 20 bushels an acre to 160 bushels an acre. In China, hybrid rice is taking over, reaching 65 percent of the crops in 2007. Quite apart from GE crops, buying new seed every year is the norm in the developed world and is becoming so in the developing world.

As for the "age-old right to save seed," Jonathan Gressel has a sharp reply:

> Many of the [GE] detractors justify their antibiotech tirades by reasoning that farmers will have to buy seed and have turned "farmer-saved seed" into a holy mantra. Almost all those well experienced in agriculture know that there is nothing worse for farming than farmer-saved seed. Yields steadily decrease because of the loss of vigor, an increase in disease, and an often massive contamination with weed seeds. Only the very best growers are chosen to grow certified seed, and even they are heavily monitored by the seed companies, who are further regulated by governmental authorities.

● The Monsanto issue is a distraction. Yes indeed, the company, under CEO Robert Shapiro in the 1980s, moved too fast, thoroughly botched the introduction of GE crops in Europe, and was secretive when it should have been transparent. Some blame it for focusing on benefits for farmers instead of consumers—yield and efficiency rather than nutrition and variety—but the customers of a seed and herbicide company are the growers, not the eaters. As for complaints that Monsanto and the other "gene giant" companies are forcing dependency on farmers, apparently there is some unclarity on the concept of "customer." The moment Monsanto's customers are unhappy with its service, they can switch to something else. Most farmers buy their seeds from a wide range of suppliers and brokers. If they want GE seeds, Monsanto is getting competition from Syngenta, Dow, and Dupont-Pioneer.

In a 1999 letter to *Science* condemning genetic engineering, Amory Lovins asked, "Is redesigning evolution to work not at its biological pace but at that of quarterly earnings reports—and to align not with biological fitness but with economic profitability (survival not of the fittest but of the fattest)—really a good idea?" Apart from his romantic notions about biological fitness and pace in agriculture, Lovins here seems to abandon the economic analysis he applies to nuclear power. He says the absence of private capital and market rewards is an argument against nuclear, but the presence of private capital and market rewards is supposed to be an argument against GE.

Being generically against corporations is no more useful than being generically against nations. There are good ones and bad ones, and sometimes good ones do bad things, and vice versa. In the case of GE opposition, the anticorporate bias is oddly selective, because GE agriculture companies are condemned, but massive multinational GE drug companies are not. The reason for that, Robert Paarlberg suggests, is that "multinational drug companies . . . deliver products with benefits widely valued in rich countries, whereas multinational seed companies do not."

Why was water fluoridation rejected by the political right and Frankenfood by the political left? The answer, I suspect, is that fluoridation came from government and genetically engineered crops from corporations. If the origins had been reversed—as they could have been—the positions would be reversed too.

There is indeed a problem with the companies that currently dominate agricultural biotech, but it's not one that environmentalists complain about, though I hope they will do so. Only a few big corporate players have survived a period of consolidation, caused partly by excessive anti-GE regulation that drove out small companies. The winners are Monsanto, Dow AgroSciences, DuPont-Pioneer, Syngenta in Switzerland, and Bayer CropScience and BASF Plant Science in Germany. This oligopoly was formed, geneticist Pamela Ronald points out, partly to aggregate and control the intellectual property of patented genes and techniques that are the engine of agribiotech. "What this means," she writes,

> is that the private companies now have even more control over who uses the technology of genetic engineering. If a particular aspect of the technology is key to the entire process—say for example, the means to introduce a gene into a plant—denial of access to a single technological component is essentially equivalent to denial of access to the entire process. This "exclusive licensing" by universities of key aspects of GE technology to private corporations greatly restricts the ability of the public research sector to develop new crops using GE.

Alternatives are emerging, fortunately. Developing countries are building their own noncorporate GE programs suited to their unique agricultural needs, often with funding and scientific assistance from the Rockefeller Foundation, the McKnight Foundation, or the Bill and Melinda Gates Foundation, and often in open-source mode, in which the new techniques are freely shared. Richard Jefferson in Australia runs a nonprofit called Cambria that develops unrestricted GE tools. With backing from Rockefeller, his group has devised two techniques that engineer around the patent-controlled GE methodologies and free them up, particularly for use in Southeast Asia.

Pamela Ronald, at the University of California–Davis, isolated a gene that gives rice resistance to a major disease. True to her inclination, the university gave Monsanto and Pioneer the option to license the gene in certain crops grown widely in the developed world, but denied licensing rights for rice in poor countries. Furthermore, the gene was widely and freely distributed throughout the world to anyone who wanted it. Chinese scientists have now developed a GE hybrid rice carrying this gene, and the disease-resistant rice seed may soon be distributed freely to growers.

With public support, GE projects for local crops are proliferating in the developing world. A roster as of 2005, according to Joel Cohen and Jennifer Thomson, includes the following:

> In Africa, four countries (Egypt, Kenya, South Africa and Zimbabwe) are developing GM apples, cassava, cotton, cowpea, cucumber, grapes, lupin, maize, melons, pearl millet, potatoes, sorghum, soybeans, squash, strawberries, sugar cane, sweet potatoes, tomatoes, watermelons and wheat. The traits included agronomic properties, bacterial resistance, fungal resistance, herbicide tolerance, insect resistance, product quality and virus resistance.
>
> Similarly in Asia, seven countries (China, India, Indonesia, Malaysia, Pakistan, Philippines and Thailand) are developing GM crops including bananas and plantains, cabbage, cacao, cassava, cauliflower, chickpeas, chilli, citrus, coffee, cotton, eggplant, groundnuts, maize, mangoes, melons, mung beans, mustard/rapeseed, palms, papayas, potatoes, rice, shallots, soybeans, sugar cane, sweet potatoes and tomatoes.
>
> In Latin America, four countries (Argentina, Brazil, Costa Rica and Mexico) are developing GM lucerne, bananas and plantains, beans, citrus, maize, papayas, potatoes, rice, soybeans, strawberries, sunflowers and wheat.

So much for the leftist dread of centralized corporate control of global food production through genetic engineering. GE instead is proving to be a tool of regional empowerment, enhanced cultural variety in foods, and the ability of farmers to sell to global markets without being controlled by them.

✦

Genetic-engineering enthusiasts have their own favorite stories that appear repeatedly in books with titles like *Seeds for the Future*, *Mendel in the Kitchen, Liberation Biology, Genetically Modified Planet,* and *Tomorrow's Table.* Here's one they all tell you that you never see in the anti-GE articles and books.

EVERYONE LOVES GE PAPAYAS IN HAWAII

The story: In the 1960s, an outbreak of papaya ringspot virus wiped out the industry on Oahu, so the growers moved to the Puna area on the Big Island, where the virus hadn't yet invaded. Then in 1992, just when the annual papaya crop reached 53 million pounds, the ringspot virus was detected in Puna. Fortuitously, that same year, field trials showed that a transgenic line of papayas incorporating a gene from the virus worked like a vaccination against the disease. The race was on. As they watched their papaya trees go barren with the virus, the growers collaborated with state officials and the developer of the GE papayas, Hawaiian virologist Dennis Gonsalves, to get regulatory approval for the resistant line. They got it through the U.S. Department of Agriculture in 1996, the Environmental Protection Agency and the Food and Drug Administration in 1997. (Conventionally bred plants, remember, face no such gauntlet.) In 1998 the new seeds were distributed free to growers. By 2001 the papaya crop was heading back to full strength. The two GE varieties, SunUp and Rainbow, were delicious, delighting consumers throughout the United States, Canada, and eventually Japan. Probably you've eaten some, because 90 percent of all Hawaiian papayas are transgenic now.

The rest of the story: European importers of gourmet foods wanted the Hawaiian GE papayas but were not allowed to bring them in. Organic papaya growers in Hawaii cleverly plant their non-GE papaya trees in the middle of GE orchards and thus are protected from the virus.

Greenpeace activists in Thailand, defying the interests of poor farmers, persuaded the government to ban field tests of GE virus-resistant papayas, but the technology is being pursued actively in China, and many expect that once China adopts GE papayas, all of Asia will. (In 2008 China inaugurated a $3.5 billion program to accelerate a "transgenic green revolution" and advance "from high-input and extensive cultivation to high-tech and intensive cultivation.")

The Hawaiian papaya story may exemplify how GE crops can best be introduced in much of the developing world. Papayas are considered a minor crop (though they are far from minor to Hawaiians and to poor people throughout the tropics), so multinational corporations had no role in the drama in Hawaii, and neither did multinational environmental organizations. The process of getting the GE papaya lines deregulated was run mainly by the growers themselves, which made the whole campaign transparent, fast, and inexpensive. Important assistance came from the public sector: Agricultural programs at the University of Hawaii and Cornell University collaborated in the effort, and the U.S. Department of Agriculture helped out with $60,000. Once the new crops went on sale, local consumers instantly relished the GE papayas, and that attitude spread to export markets overseas. It's a success story with no shadows, unlike the next one.

GOLDEN RICE SAVES LIVES, PREVENTS BLINDNESS IN MILLIONS

The story: Rice feeds half of the world's population, but it lacks important micronutrients such as beta-carotene, the precursor for vitamin A; and vitamin A deficiency is a major affliction of the world's poor. According to the World Health Organization, "An estimated 250,000 to 500,000 vitamin A–deficient children become blind every year, half of them dying within 12 months of losing their sight." The UN Children's Fund reported in 2004, "Vitamin A deficiency is compromising the immune systems of approximately 40 percent of the developing world's under-fives and leading to the early deaths of an estimated one million young children each year."

In 1992, following a Rockefeller-funded meeting on the subject, Ingo Potrykus in Switzerland and Peter Beyer in Germany decided to collaborate on finding a way to fortify rice genetically to provide the missing vitamin A. It took them seven years. Potrykus was so harassed by anti-GE activists that the Swiss government had to build him a grenade-proof greenhouse.

In 1999 the scientists sent their paper to *Nature*, describing how adding two genes to the rice from a daffodil (which also lent a yellow hue), along with one bacterial gene, did the trick, but *Nature* refused even to send the paper out for comment. Botanist Peter Raven got word of the situation and arranged for the paper to be published in *Science*, where it inspired instant acclaim as great science and a humanitarian breakthrough. In July 2000, *Time* magazine put Potrykus on the cover, with the headline, "This rice could save a million kids a year." Golden rice was a hit.

The response from anti-GE organizations was savage. "Hoax." "Fool's gold." "Trojan horse." "Deliberate deception." "Technical failure." "Useless application." "Threat to biodiversity." "Rip-off of the public trust." "Will clash with traditions associated with white rice." "Could lead to permanent brain damage." Even: "GE rice could, if introduced on a large scale, exacerbate malnutrition and undermine food security because it encourages a diet based on a single industrial staple food"—that one from Greenpeace in 2005. Egyptian scientist Ismail Serageldin spoke for many appalled scientists when he responded, "I ask opponents of biotechnology, do you want two to three million children a year to go blind and one million to die of vitamin A deficiency, just because you object to the way Golden Rice was created?" (Serageldin was then director of the Consultative Group on International Agricultural Research; currently he is head of the new Library of Alexandria.)

The rest of the story: The critics were right about one thing (and only one thing). Golden rice supplied enough beta-carotene to provide just a fifth of the recommended daily allowance of Vitamin A. Over Potrykus's strong objection, the Swiss corporation Syngenta entered the fray. Scientists there replaced one of the daffodil genes with a maize gene and got a twentyfold increase of beta-carotene in "golden rice 2," solving the vitamin A sufficiency problem. Then Adrian Dubock, a British scientist-diplomat at Syngenta, freed golden rice from its maze of patent violations involving other companies and arranged for the developing-world rights for golden rice 2 to be managed by the Humanitarian Golden Rice Network, chaired by Potrykus. Any farmer making less than $10,000 a year could get the seeds for free and own the right to breed and sow them year after year.

By 2007 field trials of golden rice were being conducted in the Philippines by the International Rice Research Institute, aided by a $20 million grant from the Gates Foundation, with the goal of freeing the GE rice for public use by 2011. The Gates Foundation also funded Peter Beyer to head the international ProVitaMinRice Consortium, which aims to "stack multiple micronutrient and bioavailability traits into Golden Rice." The next-generation rice will have increased protein, vitamin E, iron, and zinc. A really ambitious project at the International Rice Research Institute is to convert rice from a C3 plant to a C4 plant—from the low-efficiency photosynthesis mode of wheat and potatoes to the more highly evolved, higher-efficiency mode of corn and sugarcane. C4 rice would need far less water and fertilizer yet would provide a 50 percent increase in yield. "That's just the kind of long-term, high-payoff research that governments should be funding," says Philip Pardey, an agricultural economist at the University of Minnesota.

Anti-GE environmentalists fought so viciously against Golden Rice because they knew it would be the first of a cornucopia of food plants bio-fortified for high nutrition that would be widely desirable. They were right about that part.

GE CORN REJECTED AS "POISON" IN STARVING ZAMBIA

The story: In 2001 and 2002, a severe drought in southern Africa threatened the lives of 15 million people in seven countries. A 15,000-ton aid shipment of U.S. corn (about one-third GE) from the UN World Food Programme was turned away by the government of Zimbabwe on the grounds that some GE corn kernels might be planted rather than eaten, and that would endanger the country's exports to GE-averse Europe. The United States offered to grind the corn to meal so it could not be planted. Meanwhile, part of the shipment was diverted to Zambia, just to the north, where 3 million were facing famine. Zambia had accepted and eaten such shipments for six years, but this time it was rejected. "Simply because my people are hungry, that is no justification to give them poison, to give them food that is intrinsically dangerous to their health," President Levy Mwanawasa declared. "We would rather starve than get something toxic."

Outside a locked warehouse in Shimabala, Zambia, where the corn was stored for free distribution, the *Los Angeles Times* reported, an elderly blind man pleaded with officials to release the corn: "Please give us the food. We don't care if it is poisonous because we are dying anyway." In desperation, rural Zambians were eating "leaves, twigs and even poisonous berries and nuts," said the *Times*. The World Health Organization estimated that 35,000 Zambians would starve to death in the coming months. Shipments of identical U.S. corn were accepted that year without incident in Lesotho, Malawi, and Swaziland; Zimbabwe and Mozambique accepted the corn in meal form.

The lethal change of policy in Zambia was the result of a concerted effort by Europe-based environmental organizations to frighten African nations about GE crops. South Africa had already adopted GE cotton, soybeans, and white maize—a favorite food locally—but other nations were susceptible to pressure. The leaders of the Africa campaign were Greenpeace International and Friends of the Earth International, both based in Amsterdam. Greenpeace, with chapters in forty countries, had a thousand full-time staff members, and Friends of the Earth had chapters in sixty-eight countries and 1,200 full-time staff. You can find thorough documentation of the players, techniques, and effectiveness of the campaign in Robert Paarlberg's book, *Starved for Science*. Decision makers in Zambia and elsewhere were persuaded that GE crops would cause allergies, would infect their digestive tracts, would spread HIV/AIDS, would contain pig genes, and would deny them any possibility of selling their crops to European markets.

Starvation was treated as a measure of commitment to the cause. In the service of what was thought to be a higher good, the environmental movement went sociopathic in Africa. In a panel discussion in Johannesburg, Bill Moyers asked the Indian antiglobalist Vandana Shiva about the situation in Zambia. She said:

> When the same situation happened in India, with the cyclone—30,000 people dead and many hungry—when we tested the food and found it to be GM, and we just gave the information to the people who were victims, who were hungry, they led a

protest to the aid agencies and they said just because we are poor, just because we are in emergency, doesn't mean you can force us to eat what we don't want to eat. Emergency cannot be used as a market opportunity.

I propose that anyone who encourages other people to starve on principle should do some of the starving themselves. I can attest that starving just a little bit, just for a week, concentrates the mind wonderfully. Bertolt Brecht stated the operative rule: "Grub first, *then* ethics."

Just as it's worth knowing and remembering who was CEO of Exxon Mobil when it spent millions trying to discredit climate change (Lee Raymond), it's worth knowing and remembering who was leading Greenpeace International (Thilo Bode, then Gerd Leipold) and Friends of the Earth International (Ricardo Navarro) when those two organizations went to great lengths to persuade Africans that, in the service of ideology, starvation was good for them. On their watch and among their many other beneficial campaigns, their organizations—and the European nations and humanitarian NGOs they influenced—screwed up royally in Africa.

The Kenyan plant pathologist Florence Wambugu said as much in testimony to the U.S. Congress in 2003: "The primary accomplishment of the mainly European antibiotech lobby, through gross misinformation and political maneuvering, was only to keep safe and nutritious food out of the hands of starving people. . . . The antibiotech lobby asserts that the continent needs to be protected from big multinational biotech companies. This often Eurocentric view is founded on two premises: that Africa has no expertise to make an informed decision and that the continent should focus on organic farming." Dr. Wambugu went on to spell out how corporations, as well as NGOs, need to respect African autonomy:

Consumers need to be informed of the pros and cons of various agricultural biotechnology packages, the dangers of using unsuitable foreign germplasm, and how to avoid the loss of local germplasm and to maintain local diversity. Other checks and balances are required to avoid patenting local germplasm and innovations by multinationals; to ensure policies on intellectual property rights and to avoid unfair competition; to prevent the monopoly buying of local seed companies; and to prevent the exploitation of local consumers and companies by foreign multinationals. Field trials need to be done locally, in Africa, to establish environmental safety under tropical conditions.

• The rest of the story: Africa has a multitude of agricultural problems to solve; only some can be helped by transgenic technology. The core problems are malnutrition and undernutrition, both still on the increase. According to one 2008 report, "Deficiencies in macronutrients, protein, and energy, as well as micronutrients, iron, vitamin A, zinc, and iodine [are] the underlying cause of half of all child mortality." GE can help with that one—more food and better food. African small farms are rain fed (5 percent of agriculture in Africa has irrigation, versus 60 percent in Asia) and so are utterly dependent on the weather. Whole crops are lost and whole regions starve when the rains don't come. Though drought-tolerant GE crops will help somewhat, the main needs are for irrigation systems, wells, electricity to run them, and roads for transporting farm equipment and produce heading to market.

African soil is seriously degraded, in part because crop residues are used for fuel and building materials rather than returned to the soil, in part because synthetic fertilizers often aren't available. GE is no help here. Soil enrichment methods are well established. They include (to quote a Gates-funded National Research Council report):

> controlled grazing, mulching with organic matter, applying manure and biosolids, use of cover crops in the rotation cycle, agroforestry, contour farming, hedgerows, terracing, plastic mulch for erosion control, no-till or conservation tillage, retention of crop residue, appropriate use of water and irrigation, and the use of integrated nutrient management, including the judicious use of chemical fertilizers. Land-use planning and land-tenure reform are policy tools to accompany those techniques.

Africa has particularly horrendous pests. Tsetse flies torture the livestock, parasitic weeds such as *Striga* (witchweed) attack everything that grows, a new version of wheat rust from Uganda now threatens wheat crops world-wide, and flocks of millions of the red-billed quelea devour entire harvests of sorghum, keeping generations of children out of school to chase the birds from the fields. GE can help every one of these.

The point is this: Agriculture in Africa south of the Sahara is mostly tropical. Farm practices and germplasms and corporations and political stances developed in the temperate north don't much apply. "The tropics are sun rich and water poor, while temperate zones are water rich and sun poor," says plant biologist Deborah Delmer. "Most agriculture is devel-oped for the temperate zones. Most people are in the tropics. Tropical pests aren't killed by winter. Farms in the tropics have many more crops than temperate farms. Each region in the tropics should have its own research infrastructure."

• No thanks to decades of European interference, Africa is making up its own mind about the uses of biotech for its unique agricultural situa-tion. At the 2001 World Economic Forum meeting at Davos, Switzerland, physicist-essayist Freeman Dyson watched a panel debating GE crops. His report:

> It was a debate between Europe and Africa. The Europeans oppose GM food with religious zeal. They say it is destroying the balance of nature, with unacceptable risks to human health and natural ecology. They talked a great deal about a rule called the Precautionary Principle. The Precautionary Principle says that if some course of action carries even a remote chance of irreparable damage to the ecology, then you shouldn't do it, no matter how great the possible advantages of the action may be. You are not allowed to balance costs against benefits when deciding what to do. The Precautionary Principle gives the Europeans a firm philosophical basis for saying no to GM food.

In response, the Africans pointed out that the Precautionary Principle can just as well be used as a philosophical basis for saying yes. The growing population and general impoverishment of Africa are already causing irreparable damage to the ecology, and saying no to GM food will only make the irreparable damage worse. The European pretense of allowing no risk of irreparable damage makes no sense in the real world. In the real world there are risks of irreparable damage no matter what you do. There is no escape from balancing one risk against another. The Africans need GM crops in order to survive. In most of Africa, soils are poor, droughts are devastating, and many crops are lost to disease and pests. GM crops can make the difference between starving and surviving for subsistence farmers, between prosperity and ruin for cash farmers. Africans need to sell products to Europe. The European ban on GM food protects European farmers and hurts the Africans. As the Africans see it, the European ban on GM food is motivated more by economic advantage than by philosophical purity.

Theories abound on why Europe rejected genetic engineering while America accepted it. Many blame the late-1990s outbreak in Europe of mad cow disease (which has nothing to do with GE), with the resultant horror of dangers that might be hidden in food, along with distrust of government officials overeager to assuage fears that turned out to be legitimate. Americans read about that melodrama from afar. Robert Paarlberg thinks it's the differing legal and political frameworks: "The American legal system tends to use civil litigation after the fact rather than preemptive regulation before the fact to ensure consumer and environmental safety. And . . . America's two-party political system gives less space for Green Party candidates to gain election and then join governing coalitions to advocate against GMOs."

Genetic engineering has entered that special domain, long occupied by animal-rights activists and antiabortion activists, where violence is deemed justifiable. Vandalism of GE research crops and facilities, along with intimidation of researchers, is even more common in Europe than in the United States, where the FBI estimates that just one group, the Earth Liberation Front, made six hundred attacks causing $43 million in damage between 1996 and 2004. In his book *The March of Unreason* (2006), Dick Taverne examines how tortuous the rationales sometimes become: "In Germany . . . extreme [GE] opponents fire-bombed one of the Max Planck Institutes because it was conducting genetic research on petunias. They argued that as genetic modification was bound to lead to eugenics, and as this had been practised by the Nazis, such research was bound to lead to Nazism."

At the other end of the conceptual-stretch spectrum, Switzerland has a gene technology law, passed in 2004, which enforces protection of "the dignity of plants." All biotech research applications must have a paragraph spelling out how the dignity question will be dealt with. Scientists who asked for specifics were told by the ethics committee that, for example, genetic engineering must not cause the plants to "lose their independence," by which the committee meant their ability to reproduce. The geneticists inquired: Did that mean no seedless fruits and no male-sterile hybrids, both common in agriculture?

● I think the main element that distinguishes Europe from America and other parts of the world in regard to GE crops is the seriousness with which Europeans take what is called the precautionary principle. It was invoked in the Davos debate; it was invoked in the Zambia debacle; and it has had regulatory force in the European Union since 1992 and in the Cartagena Protocol on Biosafety, governing international movement of GE organisms, since 2000. As Robert Paarlberg points out, "Europe's precautionary principle had honorable origins. It first emerged in the context of a serious and well-documented environmental harm in Germany known as forest death. The German government responded with a 1974 clean air act that allowed action to be taken against potentially damaging chemicals even in the absence of scientific certainty regarding their contribution to the harm. In 1984 this same principle was then embraced for managing ocean pollution in the North Sea, another documented harm." But as time went by, evidence of harm disappeared as a precautionary principle trigger, and science was explicitly devalued.

There are a number of versions of the precautionary principle. The clearest and most often cited came out of a meeting of environmentalists in Wisconsin in 1998. Called the Wingspread Statement, it goes:

> When an activity raises threats of harm to human health or the environment, precautionary measures should be taken even if some cause and effect relationships are not fully established scientifically. In this context the proponent of an activity, rather than the public, should bear the burden of proof.

They had me at "precautionary," worried me at "some cause and effect," and lost me at "fully established scientifically." That is an illusory, unattainable goal. Nothing is fully established scientifically, ever—not gravity, not Darwinian evolution, not the safety of peanut-butter-&-jelly sandwiches. Science is a perpetual argument. More useful wording would be something like "precautionary measures should be taken during early stages while the preponderance and trend of relevant scientific evidence becomes established, and then the measures should respond to that evidence."

As Dyson noted, the precautionary principle, as currently applied, is deliberately one-sided, a rejection of what is called risk balancing. The convener of the Wingspread gathering, Carolyn Raffensperger, is widely quoted as saying, "Risk assessment embodies the idea that we can measure and manage or control risk and harm—and we can decide that some risk is acceptable. The Precautionary Principle is a very different idea that says that as an ethical matter, we are going to prevent all the harm we can." Net-benefit analysis is ruled out.

One consequence of the precautionary principle is that, in practice, it can be self-canceling. It says to wait for the results of further research, but it declares that the research is too dangerous to do. Under the banner of the precautionary principle, activists burn the fields where GE research is going on and threaten the researchers. "All technology should be assumed guilty until proven innocent," said Dave Brower, founder of Friends of the

Earth. That is a formula for paralysis. (I can imagine Dave responding, "A little paralysis might do a world of good about now.")

● Hear now "The Fable of the Steak Knives," as told by the founder of Wikipedia, Jimmy Wales. His software engineers were spending a lot of their time imagining problems that would occur on Wikipedia and then devising software solutions to head off the problems. He explained why that is the wrong approach:

> You want to design a restaurant, and you think to yourself, "Well, in this restaurant we're going to be serving steak. And since we're going to be serving steak, we're going to have steak knives, and since we're going to have steak knives, people might stab each other. How do we solve this problem? We're going to have to build cages around each table to make sure no one stabs each other.
>
> This makes for a bad society. . . . When you try to prevent people from doing bad things, the very obvious side effect is that you prevent them from doing good things.

The astronomical success of Wikipedia comes from its principle of not trying to solve imaginary problems but instead putting all of the community's effort into close attention to what actually goes on, noting genuine problems as they emerge, and then solving them as locally as possible with speed and efficiency. The whole system is success driven rather than problem driven.

Expected benefits from any act are finite and known: "Golden rice will prevent blindness in children." Imagined problems are infinite and unknown: "Golden rice might cause poor people to stop eating green vegetables; it might lead to excess vitamin A consumption; it might be a Trojan horse for corporate takeover; it might cause who knows what problems!" The apparent imbalance is treated as a contest: small and unlikely good versus large and certain harm. In this formulation, no good surprises are possible, all bad surprises are probable, and intended consequences are never what actually happen. In reality, intended consequences are what usually happen, surprises are balanced between good and bad, and they're easy to recognize and to expand on or correct, as needed.

If cellphones had been subject to the precautionary principle, the arguments against them would have included: They'll microwave your brain; they'll exacerbate the Digital Divide; they'll lead to the corporate takeover of all communications; they'll homogenize society—prove they won't! In reality none of those things occurred—though of course other problems did, such as incompatible standards and new forms of discourtesy. The main outcome was enormous, rapid success, with a vast empowering of individuals everywhere, especially the poor.

The late Mary Douglas, anthropologist and lifelong student of risk, noted that sectarian groups such as some environmental organizations separate themselves from the world with infinite demands. For them, she wrote, "there can never be sufficient holiness or safety." As a Brit, she also wondered, "What are Americans afraid of? Nothing much, really, except the food they eat, the water they drink, the air they breathe, the land they live on, and the energy they use." The economist Paul Romer adds a global perspective: "Even if one society loses its nerve, there'll be new entrants who can take up the torch and push ahead."

● The precautionary principle has been so widely recognized as a barrier to progress that, according to England's *Prospect* magazine, in 2006, the House of Commons select committee on science and technology recommended that the term "should not be used and should 'cease to be included in policy guidance.' " Various attempts have been made to draft a substitute—the proactive principle (Max More and Kevin Kelly), the precautionary approach (Nuffield Council on Bioethics), the reversibility principle (Jamais Cascio), and the anti-catastrophe principle—that one from an excellent book, *Laws of Fear: Beyond the Precautionary Principle* (2005), by behavioral economist Cass Sunstein, who now heads Obama's Office of Information and Regulatory Affairs.

I would not replace the precautionary principle. Its name and founding idea are too good to lose. But I would shift its bias away from inaction and toward action with a supplement—the vigilance principle, whose entire text is: "Eternal vigilance is the price of liberty." The precautionary principle by itself seeks strictly to stop or slow new things, even in the face of urgent need. Precaution plus vigilance would seek to move quickly on new things. Viewed always in the context of potential opportunity, a new device or technique would be subjected to multidisciplinary scrutiny and then given three probationary categories for ongoing oversight: 1) provisionally unsafe until proven unsafe; 2) provisionally safe until proven safe; 3) provisionally beneficial until proven beneficial. As the evaluation grows more precise over time, public policy adjusts to match it.

When GE food crops first went public in the early 1990s, precautionary vigilance would have monitored the brave early adopters, looking for signs of harm and signs of benefit, and especially for surprises, good and bad. (A surprising benefit from Bt corn, for example, is that it reduces mycotoxin poisoning in tortilla cornmeal because less insect damage means less fungal growth.) By the end of the 1990s, vigilance of a decade's cumulative experience would have declared GE food apparently safe *so far* and apparently beneficial *so far*. Europeans would gingerly have begun buying and planting GE food crops, and anti-GE activists, while remaining suspicious, would have stopped burning GE research fields and labs.

The emphasis of the vigilance principle is on liberty, the freedom to try things. The correction for emergent problems is in ceaseless, fine-grained monitoring, which largely can be automated these days via the Internet, by collecting data from distributed high-tech sensors and vigilant cellphone-armed volunteers. (Wikipedia, for example, is an orgy of vigilance: A cluster of diligent amateur watchers and correcters actively surveil each entry, with a response time of seconds.) Managing the precautionary process in this mode consists of identifying things to watch for as a new technology unfolds. (Does golden rice actually help with malnutrition? Are there really any instances of hypervitaminosis, too much vitamin A? Can they be headed off, given how they occur?) Intelligent precaution also would charge specific agencies to keep an eye out for unexpected correlations, such as the increases in lung cancer that developed around concentrations of asbestos—they were detected in the 1930s but not acted on until the 1980s. Tens of thousands suffered and died needlessly during that lag.

The mantra for dealing with pandemics is "early detection, rapid response." The old method of waiting for news of dead nurses in remote hospitals has been replaced by active monitoring of online chatter, active monitoring of the condition of animals sold in developing-world food markets, a network of "sentinel physicians," automated bioassays, and more to come. That's the way to organize vigilance.

One also has to credit the pioneers of excess. In the 1920s, radiation was lauded for its healing properties until a millionaire golfer named Eben Byers died from drinking a thousand bottles of a popular radium potion called Radiothor. Some of my contemporaries in the 1960s took pains to prove that the danger from excessive LSD use was not brain damage or chromosome damage, as had been predicted, but personality damage. Amateurs can be counted on to discover exactly how much video gaming leads to suicide, how many carrots lead to orange-eyed delirium, how many grizzly bears you have to hug before one eats you. I have no doubt that amateurs, not corporations or governments, will be the ones demonstrating how much GE is too much, and good luck heading them off with that line about proponents bearing the burden of proof. Nor will legions of corporate lawyers building forts around gene patents have any better luck. Biotech wants to be free.

The fact is that the fastest-moving countries now with GE crops are the developing nations that have the scientific competence and confidence to stand up to excessively cautious environmentalists—China, Brazil, India, South Africa, Argentina, the Philippines. As they go, so goes the world. Foundations such as Gates, Rockefeller, and McKnight are helping to spread the technology—in locally nuanced form—to those who need it most in the poorest nations, mostly in Africa and south Asia. Bitching and moaning, Europe will drag along after.

◆

What about God? What about the retribution we invite by playing God with genetic engineering?

A version of that question was put to me by Kathy Kohm, editor of the remarkable magazine *Conservation*. "The history of engineering is marked by a trail of unintended consequences," she said. "We don't know what we don't know. How do we walk the line between hubris and humility?" I replied:

> A lot swings on what is considered news. Ever since ancient Greek drama, hubris and unintended consequences have made great theater.
>
> Intended consequences, although more common, are not news and not theater. GMOs have been tested extensively, but we never hear of the results unless something suspicious turns up. . . . One headline you will never see is "GM Crop Again Shown OK."
>
> Technology emerges from science. Then we do science on the technology. Then we know what we know. The whole process works on a necessary blend of both hubris and humility.

I admire Prince Charles, especially for his humanizing influence on the design of cities and buildings. With his usual forthrightness, he has made a clear statement about the impiety of GE: "I happen to believe that this kind of genetic modification takes mankind into realms that belong to God, and to God alone." Pope Benedict in 2006 vilified scientists who "modify the very grammar of life as planned and willed by God. . . . To take God's place, without being God, is insane arrogance, a risky and dangerous venture."

An unlikely ally of the prince and the pope is the American leftist Jeremy Rifkin, who believes that GE violates "the boundaries between the sacred and the profane" and must be banned wholesale from the world. (Among scientists who have read his work, Rifkin is regarded as America's leading nitwit. The evolutionist Stephen Jay Gould, a considerable lefty himself, described Rifkin's biotech book *Algeny* as "a cleverly constructed tract of anti-intellectual propaganda masquerading as scholarship. Among books promoted as serious intellectual statements by important thinkers, I don't think I have ever read a shoddier work.")

Then you have Bill McKibben, who listened to working climatologists for his landmark book, *The End of Nature*, but borrowed his views on genetic engineering from Rifkin. GE, he wrote, "represents the second end of nature. . . . What will it mean to come across a rabbit in the woods once genetically engineered 'rabbits' are widespread? Why would we have any more reverence or affection for such a rabbit than we would for a Coke bottle?"

There is a common sentiment among environmentalists that everything made by nature is good and everything made by man is bad. "Four legs good, two legs bad." Nature is seen as whole and therefore holy. It is inscrutable and divine, whereas we are crass; and yet it is also fragile, vulnerable to our crass depredations.

What "nature" are we talking about, exactly? You *can't* do anything against nature, if your idea of nature includes physics, chemistry, and mechanics. Abominations can be imagined but cannot be performed. Anything you can do you can only do because nature allows it. Nuclear fission is so natural it occurs geologically. Horizontal gene flow is so natural it is the norm among microbes. Apparently what people mean when they say "against Nature" is "against my understanding of Darwinian inheritance and traditional breedline agriculture." Or maybe it's not so cosmic, and what people mean by "against Nature" is "something I'm not used to yet."

In looking for guidance on ethical issues, notions of abomination don't help much. What does help is a sense of how harms and benefits are distributed. In 1999 and again in 2003, the question of genetic engineering was examined in exhaustive detail by the prestigious Nuffield Council on Bioethics, in Britain. Their conclusion: "There is a moral imperative for making GM crops readily and economically available to people in developing countries who want them."

● Most environmentalists don't seem aware of what's going on in the biosciences these days. They don't realize that their battle against GE crops is a rearguard action in a sleepy backwater of biotech. So far we've been touring Agroecology 101 and Genetics 101—textbook stuff. Now we jump to the leading edge of biology, the new discoveries and techniques that aren't in the textbooks yet. This is where alert environmentalists should hang out, looking for powerful new tools to seize and deploy for Green agendas. And, for those so inclined, whole new dimensions of things to worry about are on offer. GE crops will be left in the dustbin of outdated frets, like an old food fad: "Remember when we thought Bt corn was the end of the world?"

———

Footnotes for this chapter may be found at **www.sbnotes.com**.

6

Gene Dreams

Microbes run the world. It's that simple.

—The New Science of Metagenomics

It was microbiologist Lynn Margulis, back in the 1970s, who first instructed me on the inventiveness of microbes. Along with codeveloping the Gaia hypothesis with Jim Lovelock, she revolutionized biology with her endosymbiotic theory, which posits that cells with a nucleus (which make up eukaryotes, including us) arose from the ingenious merging of nonnucleated bacteria (known as prokaryotes). The mitochondria in our cells are alien parasites that wound up being our primary energy factories. The light-harvesting chloroplasts in plant cells evolved from endosymbiotic cyanobacteria. Without them there would be no photosynthesis, no sugar synthesis, and no human beings. Endosymbiotic merging was the inspiration for Lewis Thomas's felicitous book title, *The Lives of a Cell* (1978).

The realization is humbling: All complex life forms were invented by creatures we think of as cooties, germs. Besides being the most diverse of all creatures, bacteria, Margulis writes, "are the oldest, having had the most time to evolve to take full advantage of Earth's varied habitats, including the living environments of their fellow beings. By trading genes and acquiring new heritable traits, bacteria expand their genetic capacities—in minutes, or at most hours." The fastest, then, as well as the oldest and most diverse of life forms, bacteria also are the only ones that can claim immortality. They don't age; they split and carry on forever.

No wonder eminent biologist Edward O. Wilson, who himself has revolutionized science half a dozen times, declared in his memoir, *Naturalist* (1994), "If I could do it all over again, and relive my vision in the twenty-first century, I would be a microbial ecologist."

One of the most thrilling books I've read recently you can download for free from the National Academy of Sciences. Here are some excerpts from *The New Science of Metagenomics: Revealing the Secrets of Our Microbial Planet* (2007):

> Every process in the biosphere is touched by the seemingly endless capacity of microbes to transform the world around them. It is microbes that convert the key elements of life— carbon, nitrogen, oxygen, and sulfur—into forms accessible to all other living things. For example, although plants tend to get credit for photosynthesis, it is in fact microbes that contribute most of the photosynthetic capacity to the planet. All plants and animals have closely associated microbial communities that make necessary nutrients, metals, and vitamins available to their hosts. The billions of benign microbes that live in the human gut help us to digest food, break down toxins, and fight off disease-causing microbes. . . .
>
> The combined activities of microbial communities affect the chemistry of the entire ocean and maintain the habitability of the entire planet. . . .
>
> Microbes can "eat" rocks, "breathe" metals, transform the inorganic to the organic, and crack the toughest of chemical compounds. They achieve these amazing feats in a sort of microbial "bucket brigade"—each microbe performs its own task, and its end product becomes the starting fuel for its neighbor. . . .
>
> The ultimate goal, perhaps in sight by 2027, would be a metacommunity model that seeks to explain and predict (and retrodict) the behavior of the biosphere as though it were a

single superorganism. Such a "genomics of Gaia" would be the ultimate implementation of systems biology.

● The transformative technique that makes all of this new science suddenly possible is the shotgun sequencing of the aggregate genomes of large samples of microbes, hence *meta*genomics. Microbes were long the "dark matter" of biology because, except for a few, they couldn't be cultured in the lab. Now, with what is called functional metagenomics, you don't have to bother with the organisms; you screen millions of DNA fragments from countless microbes, looking for new proteins that the fragments generate, and that tells you what the genes are used for.

A radical career move by Craig Venter is illustrative. Having grown up as a surfer on the California coast, in 2003 he returned to the sea to decompress from leading the massive effort to sequence the human genome. Venter took his sailboat, *Sorcerer II*, to the supposedly barren Sargasso Sea, collected samples of the water, and sent them back to his lab to be analyzed, using the shotgun sequence method he had developed for the human genome. In an April 2004 paper in *Science*, Venter's team reported that a barrel of Sargasso Sea water contained 1.2 million genes new to science (doubling the previous number), indicating 1,800 new "species" of bacteria and archaea. In functional terms, an astonishing 800 of the new genes were used for sensing or harvesting sunlight. Only 200 such genes had been discovered in all of life before that. On one sailboat trip, Venter had gone from being the world's leading geneticist to one of its leading field biologists, using the same equipment for both pursuits. To the annoyance of his rivals and critics, he carried on merrily sailing around the world, making new discoveries the whole way. As he reported, "Some 85 percent of the sequences in species are unique every 200 miles. So, instead of the ocean being a giant homogeneous soup, it's actually millions and millions of microenvironments, dynamically changing."

In his talks, Venter makes microbial ubiquity graphic:

In one milliliter (one fifth of a teaspoon) of seawater, there's a million bacteria and 10 million viruses. In the air in this room—

we've been doing the air genome project—all of you just during the course of this hour will be breathing in at least 10,000 different bacteria, and maybe 100,000 viruses. . . . So you're actually exchanging DNA with your neighbors without even doing it intentionally right now. . . . This is the world of biology that we live in, that we don't see, where evolution takes place on a minute-to-minute basis. . . . The air that we breathe comes from these organisms. The future of the planet rests with these organisms. . . . If you don't like bacteria, you're on the wrong planet. This is the planet of the bacteria.

Indeed, microbes make up 80 percent of Earth's total biomass, says famed microbial taxonomist Carl Woese. Of all ocean life, 95 percent requires a 1000x microscope to see. Bacteria have been found living in profusion a mile *below* the bottom of the seafloor, possibly as old as the sediment around them—111 million years.

Look closer to home. Ninety percent of you isn't you—only a tenth of the cells in your body are human; the rest are microbes. We are a portable swamp. One program of the emerging worldwide Global Metagenomics Initiative is called the International Human Microbiome Consortium, which is busy shotgun-sequencing all of the microbial communities that share our bodily life. We humans have 18,000 distinct genes; our microbes have 3 million. We are one species; they are diverse—a thousand species in our digestive tract (a twenty-one-foot-long bioreactor running on 100 trillion microbes), another thousand in our mouth, five hundred on our skin, another five hundred in those of us with a vagina. "It is inescapable," says the *Metagenomics* book, "that we are superorganisms composed of both microbial and human parts."

What is the actual wet weight of microbes we carry around with us? Bacterial cells are much smaller than human cells—like a honeybee versus a cat, as they say. The textbook *Microbial Inhabitants of Humans* (2004) estimates the total at nearly three pounds, about the weight of our brain.

• Every few minutes, every one of the microbes in your body (and the ocean, and the soil, and the air) is defying precaution and the sacred, playing God, performing an act illegal in Europe—swapping genes around in the endless search for competitive or collaborative advantage. Profligate, totally careless genetic engineering has been standard practice for 3.5 billion years. Lynn Margulis describes the process with characteristic pith in *What Is Sex?* (1998), a book she cowrote with her son Dorion Sagan:

> Genetic engineers have borrowed, not invented, gene shuffling. . . . Bacteria are not really individuals so much as part of a single global superorganism, responding to changed environmental conditions not by speciating but by excreting and incorporating useful genes from their well-endowed neighbors and then rampantly multiplying. . . .
>
> Imagine that in a coffee house you brush up against a guy with green hair. In so doing, you acquire that part of his genetic code, along with perhaps a few more novel items. Not only can you now transmit the gene for green hair to your children, but you yourself leave the coffee shop with green hair. Bacteria indulge in this sort of casual quick-gene acquisition all the time. Imagine you are a blue-eyed person (perhaps with newly acquired green hair) who, in a swimming pool, gulps the more common gene for brown eyes. Towelling off, you pick up genes from sunflowers and pigeons. Soon the brown-eyed you is sprouting petals and flying—eventually reproducing into gliding brown-eyed, green-haired quintuplets. This fantasy is mundane reality in the world of bacteria. . . .
>
> Unlike usually useless random mutations, batches of genes taken wholesale from another organism have already proven their mettle. The difference is similar to that between a misprint, which almost always makes a text worse, and an appropriate quote, which serves a purpose.

(That statement appearing here is an example of itself. In quoting Margulis, I engineered a working section of her text into mine. The mutation approach would have yielded nn*n*onssens xq4 mztWw.)

Gene transmission comes in five forms—two of them "vertical" and familiar to us, three "horizontal" and seemingly exotic but far more common. Vertical gene transfer is sexual or asexual—offspring inherit their genomes directly from two parents via sexual recombination (as we do) or asexually from one parent via splitting, budding, spores, or unfertilized eggs. The genes travel only down through the generations, hence the term *vertical*. With horizontal gene transfer, "genes can move along a bewildering variety of routes between genomes: sliding through bridges between cellular membranes, hitchhiking inside viruses, or even getting sucked up from the environment as naked fragments," reports an article in *Science*. (Those three means of intergenomic travel are referred to as conjugation, transduction, and transformation.)

Microbes do four of the five forms of gene transfer, everything but sex. By current estimates, 80 percent of the genes in microbes traveled horizontally at some point in their past. Some genes are more mobile than others. Carl Woese describes the most mobile as "cosmopolitan genes" or "life-style genes." They can provide fast local adaptivity. Lots of genes, though, don't "take." They show up in a genome and hang around as useless baggage until they're gradually selected out.

A shocking revelation of recent research is that horizontally transferred genes "can pass between organisms that are not even of the same species, genus, sub-kingdom or kingdom of life form." (That from the online encyclopedia Citizendium.) Working chunks of DNA have been detected moving naturally between rice and millet. Parasitic plants and fungi swap genes spontaneously with their hosts. Snake DNA has turned up in gerbils. Craig Venter's lab found that inside the fruit fly's genome is the *entire* genome of a common parasitic bacteria called *Wolbachia*, and 28 of the 1,206 bacterial genes are doing something useful for the fly. That would be like finding a complete flea genome inside ours, with mysterious functions.

Speaking of us, new discoveries about viruses indicate that "taken together, virus-like genes represent a staggering 90 percent of the human genome," says a report in *New Scientist*. Most of the genes are baggage, but some turn out to have been behind such crucial evolutionary innovations as the mammalian placenta and the development of the immune system. Suspicion is growing that gene-swapping through viruses is the dominant engine of evolution. Basically free-floating gene packets a hundredth the size of bacteria, viruses come in a hundred million varieties and overwhelmingly outnumber everything else. "The rate at which viruses shuffle DNA around," says the *New Scientist* article, "suggests that life is capable of acquiring fresh new material out of the blue, and also of making dramatic leaps in the time it takes to catch a cold." The report adds, "It is looking more and more as though the biosphere is an interconnected network of continuously circulated genes—a pangenome."

It's a transgenic world.

• Thanks to horizontal gene transfer, microbes have developed astounding skills. Tiny as they are, microbes can learn. (*E. coli* anticipate and prepare for the sequence of environments they face in our intestines during digestion.) Microbes do complex quorum sensing, both within species and between species—they are in that sense multicellular. (In order to coordinate group benefits such as biofilm structure and toxin release, they signal each other through chemical autoinducers.) They make rain on purpose. (Some bacteria have a surface protein that binds water molecules into raindrops and snow; when they get stuck in the air, this characteristic gets them back down to the ground. The total of such behavior is Gaian, a global feedback between life and the atmosphere.) They can survive for hundreds of millions of years inside rock and ice. (Microbiologist Russell Vreeland, who revived bacteria that had been trapped inside salt crystals for 250 million years, postulates that geology "acts as a gene bank for microbes." Glaciers have been described as "gene popsicles.")

The byplay between microbes and humans has always been intimate. They use us to provide food; we use them to ferment food. We cull them with antibiotics; they cull us with disease. Bacteria are the major remaining form of life that threatens us—via tetanus, typhoid fever, diphtheria, syphilis, cholera, leprosy, and tuberculosis, according to the Wikipedia entry on bacteria—and they attack our crops with "leaf spot, fire blights and wilts in plants as well as Johne's disease, mastitis, salmonella and anthrax in farm animals." But we've employed them for thousands of years in the making of "fermented foods such as cheese, pickles, soy sauce, sauerkraut, vinegar, wine, and yoghurt."

Scientists in search of ways to convert the inert cellulose of plants into usable energy are studying the miraculous bioreactor in the hindgut of termites. The complex microbial community there can turn a sheet of printer paper into a half gallon of hydrogen gas. Nothing else can do that. Steven Chu, the U.S. Secretary of Energy, has said, "Either we'll genetically engineer the microorganisms from termite guts to produce more energy from biomass than they need, or we'll adapt the chemistry within the microorganisms to process the biomass ourselves."

Craig Venter is awed by the existence of microbes "that can withstand millions of rads of radiation. Their genetic code gets blown into hundreds of little pieces. They can be totally desiccated. But you drop them in water and within 12 to 24 hours they reconstruct their chromosomes exactly as they were before, and they start replicating again." In that context, he says, "the idea of panspermia, that organisms can travel through space and land in an environment such as Earth and start replicating, is not far out at all. You could potentially view evolution as a 6- to 8-billion-year event, not a 3- to 4-billion-year event, if life can travel around the universe."

That's the power of horizontal gene transfer. Life at its most creative is transgenic. No wonder human ingenuity wants to continue expanding on microbial ingenuity, proceeding from yogurt to the drug artemisinin for malaria, from wine to jet fuel.

As a biology student, I was taught to sneer at Jean-Baptiste Lamarck, whose eighteenth-century theory of evolution was based on the idea of the inheritance of acquired characters. The giraffe's neck, he proposed, got long by a parent stretching for high leaves and then passing that trait on to the calves. Such simple-mindedness, we were told, was corrected by Charles Darwin, who substituted natural selection among random inherited variations as the mechanism of evolution. Darwin's theory was based on the artificial selection that breeders practice— classic vertical gene transfer. The more we study horizontal gene transfer these days, the more Lamarckian it looks. Convenient traits are acquired all the damn time in direct response to the environment, just as Lamarck proposed.

This has led Carl Woese to propose that a "Darwinian transition" occurred a couple of billion years ago when various organisms began protecting their own gene lines jealously, biasing toward vertical and away from horizontal gene transfer. This was the beginning of what we call species, one generation of identical jellyfish or chipmunks after another. "Species formed," says Woese, "when organisms stopped treating genes from other organisms with equal importance to their own genes."

Riffing on Woese, Freeman Dyson says:

> Some cells decided it was advantageous to keep their intellectual property private. . . . Each invention only benefited the species that invented it. Everybody else had to compete separately. Evolution then went much slower for a couple of billion years. That's what I call the Darwinian interlude. Since humans came along, that has changed again. Now we're back in an epoch when genes can be horizontally transferred.

● All biologists are genomicists now, and the pace of molecular biology is accelerating at a rate beyond even what we've experienced with information technology. The lead chronicler of biotech, researcher Rob Carlson, keeps updating what are called the Carlson curves. They chart how much faster than Moore's law we're developing techniques to sequence and synthesize DNA—to read and *write* genetic code. (Moore's law states that computer capability doubles every two years.) Thanks to its accelerating technology, the medical biotech industry is growing by 15 to 20 percent a year, and agricultural biotech by 10 percent a year.

Out of nowhere has come a whole new field called *synthetic biology*. Wikipedia describes it in application terms:

> Engineers view biology as a *technology*. Synthetic Biology includes the broad redefinition and expansion of biotechnology, with the ultimate goals of being able to design and build engineered biological systems that process information, manipulate chemicals, fabricate materials and structures, pro-

duce energy, provide food, and maintain and enhance human health and our environment.

The idea is to "play Nature," to reverse-engineer the tangled genetic code of eons and "refactor" it—write fresh genetic code that is manageable, that actually does have intelligent design instead of the infinity of moronic kludges and patches that timeless evolution confers. George Church, a leading molecular geneticist at Harvard, says that biology is at last becoming "an engineering discipline, with interchangeable parts, hierarchical design, interoperable systems, specification sheets—stuff that only an engineer could love." Rob Carlson reports that the minimalist approach to genome design is paying off: "Most synthetic DNA constructs are usually composed of just a few genes, with cutting edge designs topping out at about 15 genes. Amyris Biotechnologies is using genetic circuits of this size in modified microbes to process sugar into useful compounds, including malaria drugs, jet fuel, diesel, and gasoline analogues."

In 2008 Craig Venter told an audience in San Francisco how his team took the chromosome from one kind of bacteria, implanted it into another kind, and got it to "boot up" there, totally converting the invaded organism. "This is true identity theft at the ultimate level," he said, and marveled: "This software builds its own hardware."

The Stanford bioengineer Drew Endy likes to ask his audiences if they think it would be possible "to reprogram *E. coli* to smell like wintergreen while growing but like bananas while resting." If it were possible, what would it take—five students, four months, maybe $25,000? No, in reality it took one hobbyist one day and less than $1,000 to reprogram *E. coli* to put on an aroma show. In 2001 Endy joined Tom Knight at MIT to start the BioBricks Foundation, which supplies raw materials and tools for creating and adjusting genomes. Undergraduates and hobbyists show off their genetic creations at an annual iGEM (International Genetically Engineered Machine) jamboree. The 2007 iGEM competition attracted 576 participants in fifty-four teams from nineteen countries. Projects included, to quote a report in *Slate*:

self-flavoring and self-coloring yoghurt bacteria; bacteria that mimic the behavior and properties of red blood cells; "infector detector" organisms that indicate the presence of antibiotic-resistant microbes; a virus that could potentially be used to find and kill breast cancer cells; a living two-cell mercury-detection-and-removal crew for water filtration; and microbes that change color in a pattern meant to mimic fans doing the wave at a Mexican soccer match.

By the next year, 2008, jamboree participants had more than doubled to twelve hundred, in eighty-four teams from twenty-one countries.

Freeman Dyson finds in such jamborees a direct descendent of the annual flower breeders' show in Philadelphia and reptile breeders' show in San Diego, where competitors proudly show off new roses, orchids, lizards, and snakes they've created. "I predict," he says, "that the domestication of biotechnology will dominate our lives during the next fifty years at least as much as the domestication of computers has dominated our lives during the previous fifty years." Dyson is confident that once biotech is freed from the grip of large corporations, it will no longer seem alien or controversial: "In the era of Open Source biology, the magic of genes will be available to anyone with the skill and imagination to use it."

• The few environmentalists who are paying attention seem unsure whether to join the "synbio" party, ban it, or keep ignoring it. In 2007 Jim Thomas, from the anti-GE group ETC, wrote a survey of synthetic biology titled "Extreme Genetic Engineering." It is well researched, fair, inclusive, and only moderately alarmist. It does conclude: "In keeping with the Precautionary Principle, ETC Group asserts that—at a minimum—there must be an immediate ban on environmental release of *de novo* synthetic organisms until wide societal debate and strong governance are in place." That might have worked if biotech had stayed within a few large corporations under regulatory oversight, but those days are gone. "Every day," notes Roger Brent of the Molecular Sciences Institute, "in thousands of labs worldwide, genes, mRNAs and proteins, isolated from cells or organisms, are conveyed as bits (via the Internet) or as self-replicating molecules (via Federal Express), and reintroduced into other cells or used to engineer new organisms."

Brent is one of the major proponents in the United States of open-source biotech, along with Rob Carlson, Drew Endy, George Church, and Craig Venter. All of them are acutely aware of the dangers of bioterror and have worked directly with the government agencies responsible for biosecurity. They promote transparency and widely available biotech skills as the safest as well as the most pragmatic way to deal with potential dangers, much as the spreading of computer programming skills has helped keep the Internet healthy despite countless attacks with computer viruses, worms, and other malevolence. Craig Venter argues that the time biotech was most dangerous was when it was confined to secret government bio-weapon labs in the Soviet Union and the United States.

Decades of lab work and industrial-scale bioreactors have shown that it's easy to cripple organisms so they can't survive outside the work environment, and they're feeble even when you don't try to weaken them. As for creating wild-worthy microbes on purpose, computer scientist Rudy Rucker speculates:

> I have a mental image of germ-size MIT nerds putting on gangsta clothes and venturing into alleys to try some rough stuff. And then they meet up with the homies who've been keeping it real for a billion years or so.

One benefit of the anticipated importance of synthetic biology is a growing profusion of eclectic organizations and meetings designed to include all potential stakeholders and players right from the start—bioethicists, environmental activists, biosecurity professionals, social scientists, politicians, reporters, funders, and investors, along with the bioscientists and bioengineers.

Great names the organizations have, too—SYNBIOSAFE (in Europe), SynBERC (Synthetic Biology Engineering Research Center), International Consortium for Polynucleotide Synthesis, and the Industry Association of Synthetic Biology. The extensive public discussion called for by the ETC group is in fact happening. In 2008, for example, Drew Endy invited

ETC's Jim Thomas to publicly debate with him about synthetic biology, and I got to stage the event in San Francisco. "I want to develop tools that make biology easy to engineer," said Endy. "Powerful technology in an unjust world is likely to exacerbate the injustice," said Thomas.

At about the same time, a *New York Times* reporter visiting the Synthetic Biology Working Group at MIT noticed on their to-do list: "Grow a house."

Now is the time to ask: What are the most environmentally useful things that synthetic biology could do for human food production? Do we make ever finer adjustments to existing agriculture, create new crop plants, start over with algal vats, reinvent aquaculture and mariculture around microbes instead of fish, or what? And that's just Greener food. What about Greener fuel and materials?

◆

I have a history with organic farming—more than I realized. Reading *The Omnivore's Dilemma* (2007), Michael Pollan's natural history of American agriculture, I was surprised by this passage:

> *Organic Gardening and Farming* struggled along in obscurity until 1969, when an ecstatic review in the *Whole Earth Catalog* brought it to the attention of hippies trying to figure out how to grow vegetables without patronizing the military-industrial complex. Within two years *Organic Gardening and Farming*'s circulation climbed from 400,000 to 700,000.

At Whole Earth we did indeed promote the intensely organic publications from Rodale Institute in Pennsylvania, and I got to be friends with Bob Rodale. Pollan mentions also the influence of an essay we carried by farmer-poet Wendell Berry in praise of Sir Albert Howard, whose 1940 book, *An Agricultural Testament*, laid the foundation for the organic movement. That book begins, "The maintenance of the fertility of the soil is the first condition of any permanent system of agriculture."

Sir Albert's tome, along with earlier books such as Franklin Hiram

King's *Farmers of Forty Centuries* (1911) and George Perkins Marsh's *Man and Nature* (1864), convinced me that the quality of a civilization, and its likely longevity, can be judged by the quality of its soil. Thus I'm cheered by the current proliferation of new genres of soil-centered agroecology— *organic, permaculture, polyculture, conservation agriculture, biological farming,* and *integrated farm management.* We can add to the list *transgenic crops,* if they're designed right and used right.

This is the place to introduce formally two people I've already quoted a lot—Pamela Ronald and Raoul Adamchak. Raoul teaches organic farming at the University of California–Davis; prior to that, he was a partner in a commercial organic farm called Full Belly; he used to be president of the Board of California Certified Organic Farmers. Pam is a plant geneticist. Besides being married (with kids), they are coauthors of a charming, densely informative book published in 2008, *Tomorrow's Table: Organic Farming, Genetics, and the Future of Food.* Drawing on the daily details of their professional lives, the authors make a case for treating GE crops and organic farming as convergent techniques for feeding the most people with the least harm to the land. "To meet the appetites of the world's population without drastically hurting the environment," they write, "requires a visionary new approach: combining genetic engineering and organic farming. . . . Genetic engineering can be used to develop seeds with enhanced resistance to pests and pathogens; organic farming can manage the overall spectrum of pests more effectively."

To keep his organic certification, Raoul is not allowed to use any GE seeds. In *Tomorrow's Table,* he writes:

> As an organic farmer, I want to see more farmland transitioned to organic practices and at the same time I want to use the most powerful technologies available to create an environmentally friendly, sustainable, and high-yielding farm. . . . In the same way that the introduction of genes from wild species through breeding revolutionized farmers' management of pests, so can the introduction of genes through GE revolutionize control of diseases, insects, and nematodes for which there is presently no organic solution. GE can also greatly increase our under-

standing of what is going on in plants at a molecular level. Pam
has been working for twenty years trying to understand how
plants and microbes communicate.

I should mention what Pam has been up to with GE, because she's modest
about her accomplishments. At the University of California–Davis, one
of the world's great centers of agricultural research, she runs a large lab
devoted to improving rice for the developing world. Working with scientists in Asia and at the International Rice Research Institute in the Philippines, she helped isolate from an ancient rice strain in eastern India a gene
that confers *submergence tolerance*—a way to survive floods. In India and
Bangladesh, 4 million tons of rice a year are lost to flooding, enough to
feed 30 million people. As Raoul remarked in an interview, "For about 50
years, people have been trying to develop flood-resistant rice using conventional breeding. They've failed. Today about 75 million farmers live
on less than a dollar a day in major flood zones in places like Myanmar,
Bangladesh, and India."

Using GE techniques, Pam demonstrated that a single gene called
Sub1A was sufficient to confer submergence tolerance. With the genetic
information her lab generated, breeders in the Philippines, Bangladesh,
and India used a precision breeding technique (a kind of a hybrid between
genetic engineering and conventional breeding) to introduce the submergence gene into locally adapted high-yielding rice varieties, where it
makes the plants able to "hold their breath" for two whole weeks underwater. The submersible rice has now been tested in farmers' fields (the last
stage before release for public use) in Bangladesh, India, and Laos.

In *Tomorrow's Table*, she gently offers a challenge to the organic industry: "Because our team has also created California rice varieties carrying the submergence tolerance trait, we may be able to help our local
organic rice growers and other farmers fight weeds without herbicides."
(Organic rice growers use deep water to drown the weeds. Submergence-
tolerant rice would make the technique even more effective.) Which of
America's fifty-six organic certification programs, I wonder, will be the
first to accept Pam's submersible rice? Yes, it's engineered, but the gene
in question came from another rice plant, after all, and it does kill weeds

in a natural and old-fashioned way. (Now ask yourself what if the flood-resistant gene had come from a cattail? Or a catfish? Or a cat? Or a corporation? Where does wickedness cut in? To the rice plant, to the rice farmer, to the rice eater, *it doesn't matter*.)

I asked Pam about the patent status of her flood-loving rice. She wrote:

> The Sub1 gene is in the public domain (we felt it too valuable to third world farmers to delay getting it out there or tie up the patent rights in some complicated way). The Sub1 rice variety has been trialed in farmers' fields for 3 years now and has been yielding 2–5-fold more than conventional varieties under flooded conditions. Farmers are now bulking up the seed on their farms for planting next year and for sharing with their neighbors. The Bangladeshi national breeding stations are also bulking up for free distribution. Over the next 3 years they hope to have enough to plant 2 million acres.

● Organic is prospering these days. In the United States, organic cropland quadrupled between 1992 and 2005, to 4 million acres (that's still less than 3 percent of U.S. agriculture overall). Worldwide, the total reached 76 million acres, with Australia—Australia!—and parts of Europe nearly one-third organic. Growers and vendors can charge premium prices, sometimes triple what they can get for conventionally grown crops. To the extent that the organic boom is just a marketing phenomenon, it is fragile.

I pay extra for organic food for only one reason. I don't believe it's safer or more nutritious or higher yield or necessarily tastier than conventional agriculture. I do believe it reduces the impact of synthetic fertilizers, herbicides, and pesticides on American "soils, waters, and wildlife," so my extra payment is a public service, not a private one. Others may not be so generous when the next generation of GE agriculture introduces produce that is far more nutritious, delicious, and inexpensive than the best of current organic fare. Those crops could be and should be grown organically.

What is the essence of "organic"? The standard definition is that it includes care for the soil and the ecosystems surrounding organic farms, and that it relies on biological and mechanical controls to deal with pests and on organic materials for fertilizer. Only some keepers of organic theology go further. One authoritative document from the Netherlands reads:

> On the basis of respect for the value of naturalness, genetic engineering will be rejected as being "unnatural" because it disturbs the harmony or balance of the whole, but also because the recombinant DNA constructs used are not "natural substances" but synthetic constructs (relating to the no-chemicals approach). . . . Genetic engineering does not respect the characteristic way of being ("nature") of living organisms. Genetic engineering is based on a mechanistic and not a holistic way of thinking about life. So the objections against engineering of organic agriculture go well beyond the risks of the gene technology. They also relate to the technology itself, and the human attitude towards nature it reflects.

In my opinion, that statement proves that if you torture logic clear to death, you wind up saying quite a lot less than nothing. The title of the paper is "Organic Agriculture Requires Process Rather Than Product Evaluation of Novel Breeding Techniques." Does that mean that Europeans use no organic seeds produced by radiation or chemical mutagenesis? In the marketing world, "natural" now means anything the seller wants to charge extra for or distract your attention with. "Natural American Spirit" is the name of a cigarette brand with the tag line "100% additive-free natural tobacco." It evokes American Indian identity, distributes eco-informative fact cards, uses organically grown tobacco, and commands a premium price. In 2002 Natural American Spirit was bought by the megacorporation Reynolds American. One quarter of all male deaths in the developed world (and one tenth of all female deaths) are caused by smoking tobacco, according to the World Health Organization.

• What might a GE-inclusive organic agriculture look like? Organic farmer José Baer writes, "It would be great if there were a GE service that had a plethora of genes, and a plethora of crops, and you could pick and choose the gene that you wanted to splice into a specific crop. They would create the plant for you, propagate the seed, and provide you with your custom-ordered GE plant." One can imagine organic crops biotically engineered as Rachel Carson might do it. They would be designed in detail to protect and improve the soil they grow in, to foil the specific pests and weeds that threaten them, to blend well with other organic crops and with beneficial insects, to increase carbon fixation in the soil and reduce the release of methane and nitrous oxide, to be as nutritious and delicious as science can make them, and to invite further refinement by the growers.

Along with genetic BioBricks, let there be AgriBricks to finesse crop genomes for local ecological and economic fitness. (If Monsanto throws a fit, tell them that if they're polite, you might license back to them the locally attuned tweaks you've made to their patented gene array. Pretty soon they—or some company that replaces them—will be providing you with lab equipment.)

A great boon for local economic fitness is the revival of farmers' markets like those I grew up with in Illinois in the 1940s, only even better this time around. There were 340 farmers' markets in the United States in 1970, then 1,800 in 1994, and 5,000 by 2008. Speaking from his experience selling organic walnuts and tomatoes in farmers' markets, José Baer speculates what selling GE organic food there might be like:

> I sell in four southern California farmers markets, and there I find that the people will listen to what you tell them about your operation, decide whether they agree with it, and then buy accordingly. Organic, non-organic, whatever, it's whether they like what they hear you say about your operation. History matters, relationship with employees matters, relationship with the landscape matters, and food safety matters. Price—not so much.
>
> The thing that excites me now is that there is a consumer trend to caring what our food tastes like. This could open up some really interesting avenues in GE. I think that the farmers markets would be the best market for GE crops because you

would have the chance/ability to explain yourself. I would bring
pictures of the orchard so that they could understand that there
was nothing freakish about them, and I would explain the reasons that I am doing it (financial, food safety, and environmental). I think that the public would buy into it if it was presented
straight to them.

Thanks to the new interest in taste and freshness that Baer mentions,
along with concern about fuel costs, and the kind of bioregionalism
that my friends and I have been pushing for forty years, we're seeing the
growth of the slow-food and locavore movements, more roadside produce
stands, food co-ops, and community gardens, and the creation of subscription farms—a practice adopted from Germany, Switzerland, and Japan in
which people buy shares in the costs (and risks) of a farm and in return get
weekly delivery or pickup of great food. By eliminating middlemen, the
subscription approach, also known as community-supported agriculture,
means more money and better cash flow for the farmer and better prices
for the consumer.

● To anticipate how biotech plus organic might play out in the world,
especially the developing world, the precedent to examine is what went
right and wrong with the green revolution of the 1960s and 1970s. In 1969,
just when Paul Ehrlich was making his predictions in *The Population
Bomb* about the death of millions in 1970s and 1980s from famine, the
yields from new strains of wheat, rice, and maize were taking off in India
and Pakistan, and the Philippines had already flipped from rice importer to
rice exporter. That happened because in the 1940s the Rockefeller Foundation had set out to cure world hunger with better crops and cutting-edge
agricultural practices. One of their first hires was an Iowa farm lad with a
doctorate, Norman Borlaug.

Famines in Asia were not conjectural in the mid-twentieth century. In
1943, a famine in India killed 4 million. Chinese famines between 1959
and 1961 killed 30 million.

Starting in Mexico, Borlaug and scores of farmers and other scientists began breeding high-yield varieties of wheat and corn that could grow anywhere in the developing world. The new strains would have to be non-hybrid so the farmers could grow new crops from saved seeds, and they would have to be photoperiod insensitive—meaning they would grow any time of year. A major problem of previous high-yield varieties was that they toppled over from the weight of grain; Borlaug developed sturdy semidwarf varieties that put more of their growth into grain instead of stalk and could stand up through harvest. The plants didn't have to grow tall to tower over weeds because herbicides would keep the weeds down. As the new wheat and maize were introduced to Asia, similar breakthroughs were occurring with rice in the Philippines.

Crop scientist Jonathan Gressel recalls:

> The task in wheat was especially onerous, as the chromosome carrying the dwarfing gene had yield-reducing genes closely linked to it. Crossing these away was not easy, as it requires rare chromosomal recombination (crossing over), meaning screening millions of plants in the field. The task was done and the varieties rapidly adopted by farmers in India and China. The tripled yield of Green Revolution crops led to food security in countries on the brink of war, which justified the awarding of the Nobel Peace Prize to Borlaug and colleagues. The success of the Green Revolution ran counter to the predictions of economists, sociologists, political scientists, agronomists, and the gurus from the pesticide and fertilizer industries. They were sure the populace would not be flexible enough to adopt, would not have the infrastructure, the desire or ability to pay, and on and on. It is surprising how the self-appointed experts on agriculture do not know farmers, an issue reappearing with the rapid adoption of transgenics by farmers, especially by small, resource-poor farmers, against predictions by a later generation of pseudo-experts.

By some estimates, Norman Borlaug saved more lives—perhaps a billion—than any other human in history. The famines that Ehrlich predicted never occurred, in part because Borlaug, as obsessed as Ehrlich about the dangers of overpopulation, took the approach of providing more food now and striving for lower population later. It worked. As a bonus, there was a direct environmental benefit from higher-yield crops. In 2007 Borlaug wrote that "if the global cereal yields of 1950 still prevailed in 2000, we would have needed nearly 1.2 billion more hectares [4.6 million square miles] of the same quality, instead of the 660 million hectares [2.5 million square miles] used, to achieve 2000's global harvest. Moreover, had environmentally fragile land been brought into agricultural production, the soil erosion, loss of forests and grasslands, reduction in biodiversity, and extinction of wildlife species would have been disastrous."

The environmental movement, with its customary indifference to starvation, adopted the position that the green revolution was somehow a mistake. When Norman Borlaug set about working his magic in Africa in the early 1980s, environmentalists persuaded the World Bank and the Ford and Rockefeller foundations not to fund him. (Ryoichi Sasakawa in Japan eventually did provide support, and Borlaug now has programs in twelve African countries.) Al Gore summarized the environmentalist critique in his *Earth in the Balance*:

> Although the Green Revolution produced vast growth in Third World food production, it often relied on environmentally destructive techniques: heavily subsidized fertilizers and pesticides, the extravagant use of water in poorly designed irrigation schemes, the exploitation of the short-term productivity of soils (which sometimes leads to massive soil erosion), monocultured crops (which drove out diverse indigenous strains), and accelerated overall mechanization, which often gave enormous advantages to rich farmers over poor ones.

(Gore went on to recommend a second green revolution that will focus on poor farmers and the environment, employing, among other things "new advances in plant genetics [that] make it possible to introduce 'natural' resistance to some crop diseases and predators without the heavy use of pesticides and herbicides.")

Gore's book came out in 1992. Twenty-four years earlier, in 1968, a gentleman known as "the father of the Green Revolution in India," told the Indian Science Congress:

> Intensive cultivation of land without conservation of soil fertility and soil structure would lead ultimately to the springing up of deserts. Irrigation without arrangements for drainage would result in soils getting alkaline or saline. Indiscriminate use of pesticides, fungicides and herbicides could cause adverse changes in biological balance as well as lead to an increase in the incidence of cancer and other diseases. . . . Unscientific tapping of underground water would lead to the rapid exhaustion of this wonderful capital resource left to us through ages of natural farming. The rapid replacement of numerous locally adapted varieties with one or two high-yielding strains in large contiguous areas would result in the spread of serious diseases capable of wiping out entire crops.

The speaker was plant geneticist Monkombu Sambasivan Swaminathan, one of India's most distinguished scientists, honored in 1987 with the World Food Prize. To me, M. S. Swaminathan is a prime example of the best criticism of a new technology coming from players within that technology, where knowledge is firsthand and correctives can be applied directly and quickly, while the new field is still taking shape and bad habits haven't yet set in.

Despite Swaminathan's warnings, the Indian government oversubsidized the pumping of water for irrigation, draining ancient aquifers, and many farmers overapplied pesticides and herbicides to the point where they did poison the waters and cause medical harm. Of the world's twelve worst "persistent organic pollutants," seven are pesticides, including chlordane, endrin, toxaphene, DDT, and dieldrin. All have been phased out in developed countries, but some are still used in developing countries, and they persist in the environment, causing cancer, birth defects, endocrine disruption, immune dysfunction, and, it now appears, diabetes.

From Swaminathan's sequence of positions of responsibility as head of the International Rice Research Institute; the International Union for the Conservation of Nature and Natural Resources; the World Wildlife Fund for Nature (India), and cochair of the UN Millennium Task Force on Hunger, he has pushed for what he calls the "Evergreen Revolution—food for all and forever, on an environmentally friendly and socially sustainable basis." He is a proponent of *ecotechnology*, which he defines as "technologies that are rooted in the principles of ecology, economics, gender and social equity, employment generation, and energy conservation."

In a 2006 speech, Swaminathan decried the organic-farming movement's shunning of genetic engineering and applauded "what we call 'green agriculture,' which is now becoming very popular in China. The difference between organic farming and green agriculture is: You use integrated pest management, integrated nutrient supply, scientific water management—all methods by which the production potential of the soil is not reduced—*and also* you can use molecular breeding or Mendelian breeding, whichever is most appropriate." He added, "Our ability to face the challenges of global warming and sea level rise will depend upon our ability to harmonize organic farming and the new genetics."

• Another old green-revolution hand with an insider's critique is agricultural ecologist Sir Gordon Conway. While working in Borneo in the 1960s, he became one of the pioneers of "integrated pest management." From 1998 to 2004, he was president of the Rockefeller Foundation, and during that time he wrote an important book, *The Doubly Green Revolution* (1999). It noted the shortcomings of the original green revolution (excessive water use, excessive advantage to rich farmers, neglect of soil maintenance) and proposed how to remedy them. Conway expects the doubly green agricultural revolution to expand opportunities for the poorest farmers and to emphasize conserving natural resources and the environment while using GE to increase yield yet further. "Our capacity to build ecology into the seed," he writes, "is largely a consequence of modern biotechnology." The new "gene revolution" will be more adroit than the green revolution for two reasons but harder to implement for one reason, which he is trying to fix. The advantages this time are the boon of increasingly sophisticated ecological science and the boon of GE, but, as I mentioned before, GE carries an impediment much more serious than what environmentalists worry about.

The problem is intellectual property. "Food biotechnology," Conway said in a 2003 speech,

> was introduced just as globalization changed the boundaries between public and private—from *public* good, *public* domain, *public* obligation to *private* enterprise, *private* decision-making, *private* advantage. International rules controlling the rights to private ownership of research and technology have been changed, while, at the national and international levels, governments have appeared almost passive, as if ceding basic responsibilities for the public good to the private sector. . . .
>
> Universities, particularly in the US, now license most of their scientific innovations to private companies, including important enabling technologies—the technologies for conducting further research. As a result, three-fourths of the new biotechnology products, including those originally made possible by publicly supported research, are controlled by the private sector. . . . Private interests now dominate all aspects of research, production and the marketing of biotechnology. Even the regulatory systems favor big corporations with cadres of lawyers. . . .
>
> Fierce competition and low margins in the seed industry compel companies to stockpile IP [intellectual property] that does not have sufficient market value for development, so as to keep it out of the reach of competitors. This tends also to make it unavailable to public scientists still willing to work on crops for poor farmers. The number and complexity of ownership rights that must be negotiated and paid for to take a product to market have multiplied so quickly that some useful products are sitting in greenhouses going nowhere and some useful ideas are not being pursued.

While at Rockefeller, Conway opened two formal pathways around the intellectual-property problem. The first, built in collaboration with the

McKnight Foundation, works with universities to keep their biotech IP that is licensed or patented for private use also "available for public-sector humanitarian work." (The name of that one is the Public IP Resource for Agriculture—PIPRA, based at UC Davis.) The second is a fiendishly clever partnership machine. The African Agricultural Technology Foundation (AATF), based in Kenya and led by Africans, handles nothing but information and agreements. As Conway describes it,

> It is a way of giving very poor nations the tools to determine what new technologies exist in the public and private sectors, including but not limited to biotech; which ones are most relevant to their needs; how to obtain them and how to manage them; and how to develop nationally appropriate regulatory and safety regimes within which to introduce them. . . . [It will offer] its partners access to advanced agricultural technologies that are privately owned by companies and other research institutions on a royalty-free basis. In exchange for access to these technologies, the AATF will identify partner institutions that can use them to develop new crop varieties that are needed by resource-poor farmers, conduct appropriate biosafety testing, distribute seed to resource-poor farmers, and help create local markets for excess production. Most of the major international seed companies and the US Department of Agriculture have expressed a serious interest in working with the AATF to accomplish its goals.

If that works, if the large corporations and northern governments join the effort, and if environmentalists join the effort or step out of the way or are heaved out of the way, then wondrous things are in the GE pipeline, for Africa and elsewhere.

● The first order of business is biofortified food. Cassava, a root crop, is a drought-resistant major staple for 800 million people in Africa, Latin America, and parts of Asia. It has plenty of starch but is grossly deficient in protein, vitamins, and micronutrients. The daily diet of cassava in the developing world has a third of the protein a person needs and a tenth of the vitamins, and it carries a dose of cyanide that is particularly harmful to the undernourished.

In 2005 a project called BioCassava Plus, funded by the Bill and Melinda Gates Foundation, undertook to engineer a radically improved cassava. It had eight goals for the new cultivar. In terms of nutrition, a daily diet should provide all a person needs of bioavailable protein, vitamin A, vitamin E, iron, and zinc. In addition, the new cassava should be free of cyanide, should be storable for two weeks instead of one day, and should be resistant to the viruses that afflict the crop. Each trait would be engineered separately and then stacked into a single all-purpose crop plant. "This is the most ambitious plant genetic engineering project ever attempted," says the project head, plant biologist Richard Sayre from Ohio State. "One advantage of transgenics is that it's fast when it works, so we can get a product in one year. . . . When all these traits get stacked into what will be a farmer-preferred cultivar from Africa, this work will be done by African scientists in African laboratories. We're developing the tools mostly in the United States and Europe, but once those tools are in place, it becomes an African-owned and developed project." Field trials have begun in Kenya and Nigeria.

Along with golden rice, the BioCassava project is leading the way for what is called the second generation of GE innovation. The first generation—Bt corn and Roundup Ready soybeans and such—focused entirely on farm productivity and paid its way quickly. Over eighty GE crops, tested in over 25,000 field trials, proved the safety and efficacy of the technology. Now, building on the lessons from that work, the second generation of GE aims straight at the consumer to provide nutritious, delicious food, free of allergens and toxicity, that anybody can grow.

Another venture of the Gates Foundation is the African Biofortified Sorghum Project, with Florence Wambugu's Africa Harvest Biotech Foundation leading a consortium of nine institutions, including DuPont-Pioneer. Sorghum is a drought-tolerant staple for 500 million worldwide. The GE version will improve digestibility and add vitamins A and E, iron and zinc, and three amino acids. Greenhouse trials are under way in South Africa. (Vitamin A, incidentally, is currently distributed to the developing world in the form of 500 million capsules, costing about a dollar apiece. Getting the same amount of vitamin A from a fortified crop will cost about a fifth of a cent.) GE bananas are also being developed to provide a full daily allowance of vitamins A and E and iron for countries, like Uganda, that rely on bananas as their major food source.

"Greenpeace will fight to keep GE bananas, cassava, and sorghum from poor countries' fields, just as it will keep opposing golden rice, says Janet Cotter of Greenpeace's Science Unit in London." That quote was in an April 2008 issue of *Science*.

● A journalist I know, Gregg Zachary, wrote in 2008 about a little-noticed agricultural revolution going on in Africa:

> Exports of vegetables, fruits, and flowers, largely from eastern and southern Africa, now exceed $2 billion a year, up from virtually zero a quarter-century ago. . . . "The driver of agriculture is primarily urbanization," observes Steve Wiggins, a farm expert at London's Overseas Development Institute. As more people leave the African countryside, there is more land for remaining farmers, and more paying customers in the city. . . . Multinational corporations are becoming more closely involved in African agriculture, moving away from plantation-based cultivation and opting instead to enter into contracts with thousands, even hundreds of thousands, of individual farmers. China and India, hungry to satisfy the appetites of expanding middle classes, view Africa as a potential breadbasket.
>
> A method known as "contract farming" has become a crucial instrument of African empowerment. Buyers agree to purchase everything a farmer grows—coffee, cotton, even fish—freeing him from the specter of rotting crops and allowing him to produce as much as possible. And because the buyers—some of them domestic companies, others multi nationals—profit, they have a stake in farmer productivity and

> an incentive to provide such things as training and discounted seeds. . . . International buyers of major African crops from Europe, Asia, and the United States have told me repeatedly that small farmers in Africa, relying on their own land and family labor and using few costly inputs such as chemical fertilizers, are more efficient producers than plantations.

My bet is that, as with cellphones, much of the innovation in GE foods will take place in the developing world. If the organic food industry in the North continues to ban everything transgenic, it may find that it loses market and cachet to GE foods that are better tasting, better for you, and kinder to soil and ecology.

The leading edge in America, I suspect, will be GE foods that offer benefits seen as somehow medical. Coming soon is a GE pig whose pork has heart-healthy omega-3 fatty acids as good as those found in fish. Healthful bacon! (The operative gene comes from a roundworm, for those keeping score.) Now that we know that resveratrol in red wine is what keeps the French living longer despite all that butter, there's a GE wine on the way from China that has six times as much resveratrol. (The relevant gene was engineered in from a wild vine mainly for fungus resistance, which is what resveratrol is originally for, from the grapevine's perspective.) Researchers in Texas have developed a GE carrot that carries enough calcium to head off osteoporosis for people who can't get their calcium from dairy products. And DuPont-Pioneer is bringing out a GE high-oleic soybean oil they call TREUS that eliminates trans fats from cooking—more good news for American hearts. (No exotic genes in this one; the GE technique "silences" a gene so that the soybean makes monounsaturated oleic acid—as in olive oil—instead of less healthful polyunsaturated linoleic acid. It's an interesting borderline case. Will the anti-GE crowd still protest even though there are no transgenes involved in this engineered food from a GE corporation?)

I presume that few will complain as genetic engineering is increasingly deployed to head off disease in humans. There are half a dozen projects under way to disable or sterilize mosquitoes so that they cannot transmit malaria and dengue fever. Japanese scientists are developing a strain of rice that delivers cholera vaccine; Korean scientists are working on a form of tomato with a vaccine against Alzheimer's. A dental scientist in Florida has devised a permanent cure for cavities with an altered version of the bacterium that causes them, dear old *Streptococcus mutans*. And mad cow disease can be eliminated totally from livestock (and thus from us) with a combination of GE and cloning. I wonder how that will play in Europe, where the 1990s mad cow outbreak is what set many people against GE in the first place.

✦

Next come trees. Decriers of one thing or another often declare, "Our children will never forgive us if we fail to *blah blah blah*." In reality I find that later generations don't look back much; if they do look back, they don't notice whatever the issue was; if they do notice the issue, they don't care about it. I grew up with an exception to that. In Michigan, where I spent all my young summers, my great-grandparents' generation had clear-cut *the whole state* of its vast forests of white pine and Norway pine. "Daylight in the swamp, boys!" My parents' cottage was in one of the two remaining groves of virgin pines in the whole Lower Peninsula; we had white pines 150 feet high. I knew what was gone, and forgiveness was not in me. You can visit a sample of the generational wrath in a fine novel called *True North*, by Jim Harrison. The son of one of the murderers of the northern Michigan forest misspends his life in study and loathing of his father's crimes against the land.

It's not just in the past. A couple of years ago, my wife and I visited Tasmania, in part to see the tallest hardwood trees in the world, the lordly *Eucalyptus regnans*, nearly 300 feet high. Some are protected. Many are still being clear-cut, not for lumber but to be chipped and processed into cardboard. It's like seeing a Stradivarius broken up to start a fire.

So here's what I want. I want commercial wood grown directly. Let it be engineered as clean low-lignin pulpwood, or as timber so straight-grained and close-grained and beautiful and inexpensive that cutting a wild tree for lumber would seem ludicrous. Such trees are best concentrated in plantations, easily harvested, leaving more forests to be wild.

Throughout the world, temperate and boreal forests have been coming back since 1950 thanks to the increase in tree plantations. In a *Foreign Affairs* article titled "Restoring the Forests," David Victor and Jesse Ausubel wrote: "An industry that draws from planted forests rather than cutting from the wild will disturb only one-fifth or less of the area for the same volume of wood. Instead of logging half the world's forests, humanity can leave almost 90 percent of them minimally disturbed." The authors elaborated:

> According to the UN Food and Agriculture Organization (FAO), one-quarter of industrial wood already comes from such farms, and the share is poised to soar once recently planted forests mature. At likely planting rates, at least one billion cubic meters of wood—half the world's supply—could come from plantations by the year 2050. Semi-natural forests—for example, those that regenerate naturally but are thinned for higher yield—could supply most of the rest. Small-scale traditional "community forestry" could also deliver a small fraction of industrial wood. Such arrangements, in which forest dwellers, often indigenous peoples, earn revenue from commercial timber, can provide essential protection to woodlands and their inhabitants.

Environmentalists have done a great job establishing and promoting the leading sustainable forestry certification program, the Forest Stewardship Council. Look for the FSC logo when you buy lumber. Of the 140,000 square miles of sustainably logged forest currently approved by the FSC, about a quarter is in plantations. However, because plantations are where genetically engineered trees are being introduced, the FSC will unfortunately not be helpful in distinguishing the best sustainable practices in that part of the industry: FSC certification excludes "wood harvested from areas where genetically modified trees are planted." Maybe some of the more adventurous FSC professionals could bud off a subsidiary called the FfSC—Frankenforest Stewardship Council—to oversee and evaluate the arrival of GE tree plantations to make sure they are maximally Green.

• A major issue will be gene flow. The first tree being engineered by everybody—China, the United States, Britain, etc.—is the poplar, because it grows fast and grows big. Some want it for pulp, some for plywood, some for biofuel. Those growing poplars for pulp or biofuel are crafting low-lignin varieties for cheaper and cleaner processing, leading to less toxicity pouring out of the pulp mills. But it's the pest-resistant Bt poplars in Chinese plantations that will tell the most about how much of a problem gene flow will be. Having proved highly successful against insects, two Bt varieties of the European black poplar were released in 2002 for use in China's huge reforestation effort. Huoran Wang from the Chinese Academy of Forestry reported in 2004:

> The Chinese government has set a lofty target for forestry development: that forest coverage will reach 19 percent of the total land area by 2010 and 23 percent by 2020. . . . Forest genetics, genetic modification and domestication of forest trees will, beyond all doubt, be asked to make contributions to the goal. . . .
>
> It is estimated that one million GE *Populus nigra* [black poplar] trees have so far been propagated and used in the establishment of plantations. . . . However, the accurate area of GE plantations cannot be assessed because of the ease of propagation and marketing of GE trees and the difficulty of morphologically distinguishing GE from non-GE trees. A number of individual nursery-men at markets declare that their planting materials are GE trees produced through high-tech, for a higher price. Consequently, a lot of materials are moved from one nursery to another and it is difficult to trace them. . . .
>
> It is almost impossible to reduce the risk of gene flow from GE trees to non-GE trees through isolation distances

because of the ease of natural hybridization between poplars of the same section, and poplar trees are so widely planted in northern China that pollen and seed dispersal cannot be prevented.

There you have the very circumstance that environmentalists have most feared: transgenes loose in the world, and in a long-lived organism that outcrosses enthusiastically. It is a golden opportunity for definitive field research on gene flow—how much occurs and how much harm it does.

If thorough field research is done on China's poplars, I predict the following:

- A fast, cheap method of detecting the transgenes in poplars will be devised.
- Much less gene flow than expected will be found. (The impediment is what a paper by the Poplar Working Group in America calls "genetic inertia." Among the elements of inertia listed in the paper are "delayed flowering, tree longevity, vegetative persistence, extensive wild stands, [and] dilution of plantation-derived propagules by those from wild stands.")
- The evolution of Bt-resistant pests will occur at a manageable pace, thanks to nonengineered poplar stands functioning as refugia for unspecialized bugs.
- The harm to the ecology of nonengineered poplars from transgenes will be found to be minuscule, especially compared to the impact from climate change, habitat loss, and alien-invasive species.
- Nevertheless, some varieties of sterile genetically engineered poplars will be developed just to allay ongoing fears of transgene "contamination."

Maybe I'm wrong. The proof is in the research—unless it isn't done, and we are left with dueling assertions, which is no use at all. You've seen my assertions. Here's one on the other side from the Global Justice Ecology Project: "The inevitable contamination of native forests by genetically engineered trees may cause destruction of wildlife, depletion of fresh water and soils, collapse of native forest ecosystems, cultural genocide of forest-based indigenous communities, and serious effects on human health." I doubt that, but China will tell.

✦

One of the major drivers of GE agriculture is climate change. How we farm has to switch from being a climate problem to a climate solution. Some of that transformation will come from better practices that enrich soil and preserve wildland, some from engineered "ecology in the seed."

Consider nitrous oxide, a greenhouse gas three hundred times worse than carbon dioxide that fumes up from soil drenched with chemical fertilizer. If the use of nitrogen fertilizer went down by a third, says a report in *New Scientist*, that "would reduce greenhouse emissions by more than grounding every single aircraft in the world." More organic farming will help. So will new varieties of rice and other crops now being engineered for far more efficient uptake of nitrogen from fertilizer, so less is needed, saving the farmer money while reducing atmosphere and water pollution.

Consider also the gases that come out of us and our livestock and pets exhaling, burping, and farting. Lovelock says that accounts for 23 percent of all greenhouse-gas emissions. Australia is on the case. One project there is engineering lower-lignin grass that would reduce the methane emitted from cows by 20 percent. Another is trying to transfer the digestive bugs from methane-free kangaroo guts to cow guts for a hoped-for 15 percent methane reduction. (One anticipates a climate-friendly probiotic yogurt for humans called Roo.)

As with energy efficiency, close attention turns up all sorts of possible gains. You notice things like *plantstones*, also called *phytoliths*. They are microscopic silica balls that lock up plant carbon in the soil for thousands of years. All crop plants have them and could be genetically encouraged to have more, earning farmers carbon credits for sequestering carbon.

Every GE-capable foundation, corporation, and government is working on drought-tolerant and salt-tolerant crops, especially for climatically fragile Africa. Such crops are so crucial for adapting to climate change that they are sure to come, and the sooner the better. But ecologists are rightly worried about these crops going feral—weedy and invasive—in the dry or salty terrain they're designed to thrive in, because there is less native plant competition in such harsh land. This is a case where serious ecological research must be done during the field trials, to figure out how best to contain the newly talented crops.

Biofuels were supposed to be a carbon-efficient reducer of greenhouse gases, but converting food crops to biofuels turned out to be an economic fiasco. Second-generation biofuels rely on biotech to make fuel out of non-food plants such as switchgrass, jatropha (for its oily seed), hemp, poplars, willows, and agricultural waste such as straw, corn stover, and forestry slash. Some engineering of the plants might help to turn all that tightly bound cellulose into fuel, but most of the work will be done by vast quantities of carefully tuned microbes turning cellulosic dross into gold. "If [the microbes] are unhappy with what they are doing, they are going to evolve away from what you want them to do," warns Craig Venter. "A key part of the future is going to be designing a system where they are not grossly unhappy." With his new company, Synthetic Genomics, Venter is one of countless entrepreneurs in the biofuels gold rush.

"The fuel-and-oil industry is a multi-trillion-dollar industry," he says.

> The same oil that gets burned as fuel is also the entire basis for the petrochemical industries, so our clothing, our plastics and our pharmaceuticals all come from oil and its derivatives. . . . Right now oil is being isolated around the globe, and there is a major effort in . . . transporting that oil around to a very finite number of refineries. Biology allows us to make these same fuels in a much more distributed fashion. I envision maybe a million micro-refineries. Companies, cities and potentially even individuals could have a small refinery to make their own fuel. This would eliminate a lot of the distribution problems and associated pollution.

That's one vision. Another may be seen in a letter to the editor in England's Green paranoia magazine, *The Ecologist:* "The greatest nightmare we face involves the genetically modified organisms (GMOs) we are creating with an enhanced ability to liquefy cellulose, the basic building blocks of plants. . . . Some of these GMOs will escape, adapt and proliferate. They could become a green plague, causing a meltdown of the vegetable kingdom. This is not alarmist." (James Watson's 1977 comment about cellulase in the previous chapter should lay to rest the ever-recurrent "green goo" hypothesis. If green goo could thrive in the world, microbes would have invented it long ago. If we try to create green goo, microbes will defeat it.)

Synthetic biologist George Church has yet another vision: "The most sustainable source of energy is sunlight and the most convenient products are pipeline-compatible petrochemicals. So I would aim for a perennial plant system that secreted pure chemicals—octane, diesel, monomer for plastics, etc.—into pipes without need for further purification."

In another distributed-energy solution, Venter wants to install his happy, skillful bacteria right at the site of maximum efficiency. "My new company," he told the San Francisco audience, "has a deal with BP to try and use biology deep in the earth to stop mining coal by biologically converting that coal into methane. . . . This doesn't stop taking carbon out of the ground, but it's about a tenfold improvement over mining coal and burning it."

• I've learned something about the environmentalist mindset from having worked with the U.S. intelligence community in recent years. The national-security perspective on any new technology is a set of anxious questions. How can this new thing hurt us? Who is creating it, promoting it, or grabbing it, and what is their agenda? How might that agenda shape the technology in harmful ways?

Such technoparanoia has a way of being self-fulfilling. It institutionalizes distrust, establishing an interpretive apparatus that sees only threat and only enemies, and thereby helps to create both. Whether you're defending a nation or the natural world, a more useful assumption with any new technology is that it is neutral, and so are the people creating it and using it. Your job is to help maximize its advantages and minimize its harm. That can't be done from a distance. Particularly for environmentalists, the best way for doubters to control a questionable new technology is to embrace it, lest it remain wholly in the hands of enthusiasts who think there is nothing questionable about it.

I would love to see what a cadre of dedicated environmental scientists could do with genetic engineering. Besides assuring the kind of transparency needed for intelligent regulation, they could direct a powerful new tool at some of the most vexing problems in our field, such as heading off alien-invasive species like kudzu and mitten crabs or detecting toxins in water with a cheap bacterial sensor. (The last one has already been done by three students from Brown University; it won Best Environmental Project at the 2008 iGEM Jamboree.)

In the 1970s I saw hackers transform computers from organizational-control machines into individual-freedom machines. Where are the Green biotech hackers? The appropriate attitude was expressed by a participant in one of the new grassroots Maker Faires: "We are grabbing technology, ripping the back off of it and reaching our hands in where we are not supposed to be." Whether it's slum dwellers in India reverse-engineering cellphones to write their own repair manuals, or Drew Endy distributing BioBrick parts to amateur bioengineers, or contract farmers in West Africa reinventing traditional African polyculture around new GE crops and livestock, the attraction is grassroots empowerment.

I think every environmental organization would be well served by working closely with such people. The true nature of any new technology can be learned best from what enthusiasts do with it. Critiques based on the experience of practitioners, rather than on ideology or theory, have real bite, and corrections for problems can be tried and proved at street level before they're broadly recommended.

If environmentalists and others dubious about genetic engineering had taken this approach from the start in the 1980s, cleaving close to researchers out in the fields, they would have quickly discovered that GE foods are safe to eat, GE crops can be ecologically beneficial, and corporations are significant but not necessarily controlling players in the technology. Environmentalists would have helped lead the doubly Green agricultural revolution in the developing world instead of delaying it. Two decades of time, money, people, and credibility wasted on anti-GE activism would have been saved for solving real rather than imaginary environmental problems, and genetic technology would have arrived in the world Green from the ground up.

✦

With that, the main news items of this book are now done. Cities are Green. Nuclear energy is Green. Genetic engineering is Green. The rest of the book addresses how not to repeat the mistakes we made on those three, how deep inhabitants take care of the nature around them, and how to manage the planct's global-scale natural infrastructure with as light a touch as possible, considering.

Footnotes for this chapter may be found at **www.sbnotes.com**.

7

Romantics, Scientists, Engineers

I saw myself
a ring of bone
in the clear stream
of all of it

and vowed,
always to be open to it
that all of it
might flow through

and then heard
"ring of bone" where
ring is what a

bell does

—Lew Welch, *Ring of Bone*

Every day I wonder how many things I am dead wrong about.

—Jim Harrison, *True North*

Environmentalists own the color green. That's extraordinary, an astonishing accomplishment. No movement has owned a color globally since the Communists took over red. Red means nothing now. How long will Green mean something?

"What is the environmental movement?" It was the editor of a Green magazine asking. I heard myself say, "The environmental movement is a body of science, technology, and emotion engaged in directing public discourse, public policy, and private behavior toward ensuring the health of natural systems."

My theory is that the success of the environmental movement is driven by two powerful forces—romanticism and science—that are often in opposition, with a third force emerging. The romantics identify with natural systems; the scientists study natural systems. The romantics are moralistic, rebellious against the perceived dominant power, and dismissive of any who appear to stray from the true path. They hate to admit mistakes or change direction. The scientists are ethical rather than moralistic, rebellious against any perceived dominant paradigm, and combative against one another. For them, identifying mistakes is what science *is*, and direction change is the goal.

It's fortunate that there are so many romantics in the movement, because they are the ones who inspire the majority in most developed societies to see themselves as environmentalists. But that also means that scientists and their perceptions are always in the minority; they are easily ignored, suppressed, or demonized when their views don't fit the consensus story line.

A new set of environmental players is shifting the balance. Engineers are arriving who see any environmental problem neither as a romantic tragedy nor as a scientific puzzle but simply as something to fix. They look to the scientists for data to fix the problem with, and the scientists appreciate the engineers because new technology is what makes science go forward. The romantics distrust engineers—sometimes correctly—for their hubris and are uncomfortable with the prospect of fixing things because the essence of tragedy is that it can't be fixed.

Romantics love problems; scientists discover and analyze problems; engineers solve problems.

• That is a gross oversimplification. Stereotypes were not responsible for the burst of U.S. environmental legislation passed in the 1970s—the Clean Air, Clean Water, and Endangered Species acts, and the creation of the Environmental Protection Agency. What would I call the dedicated lawyers who got those bills written, passed, and signed—"political engineers"?

Where in my character set are the duck hunters who pioneered the conservation movement in the 1930s by protecting wetlands and who are still at it seventy years later? Some 24 million acres of North American waterfowl habitat are being preserved, protected, and restored by the 775,000 well-armed members of Ducks Unlimited.

Real people, not paper cutouts, made recycling happen, cleaned the air in Los Angeles and the Thames in London, and elevated ecology to a philosophy; made ecotourism an industry, wildlife films an entertainment genre, and *watershed* a term of art for planners; planted countless urban trees and slowed the destruction of the Amazon rain forest; stopped acid rain and ozone depletion, saved condors and whooping cranes, built global Green organizations, created wildland parks at every governmental level from county to World Heritage Sites . . . The list could fill the rest of this book. It does fill Paul Hawken's broadside on the proliferation of environmental activist groups, *Blessed Unrest: How the Largest Movement in the World Came into Being and Why No One Saw It Coming* (2007).

I'm going to stick with my stock characters, though, because they offer a way to think about some important changes that are going on. When concern about climate change went mainstream all over the world in 2007, Greens everywhere felt vindicated. "Today's torrent of environmental progress," declared the head of Sierra Club that summer, "rivals that in the heady years around the first Earth Day in 1970." The world was finally coming around to the Green point of view, and all environmentalists had to do was to seize the opportunity and bear down on their agenda to win final victory.

Wrong. The long-evolved Green agenda is suddenly outdated—too negative, too tradition-bound, too specialized, too politically one-sided for the scale of the climate problem. Far from taking a new dominant role, environmentalists risk being marginalized more than ever, with many of their deep goals and well-honed strategies irrelevant to the new tasks. Accustomed to saving natural systems from civilization, Greens now have the unfamiliar task of saving civilization from a natural system—climate dynamics.

It may seem hardest to change course when you think you're triumphant, but it's actually an opportune time. Resources abound; new people with new ideas show up. With the old guard swamped by events, Young Turks can strike out in divergent directions. An unsentimental review of the past can toss out entrenched ideas that are no longer useful and poke around in long-taboo areas for potential new value. That's the mode I'll try to frame here, not just for my fellow mossback environmentalists but also for the new climate-driven environmentalists—the Green bio-hackers, Green technophiles, Green urbanists, and Green infrastructure rebuilders.

◆

It was romantics—charismatic figures such as Henry Thoreau, John Muir, David Brower, Ed Abbey, Dave Foreman, and Julia Butterfly Hill—who taught us to be rings of bone, open to all of it, ready to redirect our lives based on our deepest connection to nature. The year I graduated from Stanford, Brower launched the Sierra Club's Exhibit Format series of nature photography books. His first one, *This is the American Earth* (1960), made with photographer Ansel Adams, set me on a path I'm still on. Desert writer Ed Abbey introduced the further romance of protest, and role models like Earth-Firster Dave Foreman and mythic tree sitter Julia Butterfly Hill played it out.

Certain knowledge of what to fight for, and what to fight against, gives meaning to life and provides its own version of discipline: *never* give up. That kind of meaning is illusory, I now believe, and blinkered. Fealty to a mystical absolute is a formula for disaster, especially in transformative times.

California was a great place to get over mysticism in the 1960s and 1970s. Such an endless parade of gurus and mystics came through, peddling their wares, that they canceled each other out. They couldn't really compete with the drugs, and the drugs canceled each other out as well. Fervent visions, shared to excess, became clanking clichés. All that was left was daily reality, with its endless negotiation, devoid of absolutes, but alive with surprises.

In 1997 my growing distrust of romanticism in all its forms was crystallized by a book: *The Idea of Decline in Western History*, by Arthur Herman, which explores one question: What is behind the ever-popular narrative of decline? Decade after decade, leading intellectuals in Europe and America explain that the world is going to hell, progress is a lie, and bad people, bad ideas, and bad institutions are to blame for the irreversible degradation of all that is true and good.

Overwhelming real-world evidence to the contrary matters not at all to the calamitists.

Herman distinguishes two forms of the lament: historical pessimism (Jacob Burckhardt, Oswald Spengler, Henry Adams, Arnold Toynbee, Paul Kennedy) and a much more frightening cultural pessimism (Arthur Schopenhauer, Friedrich Nietzsche, Martin Heidegger, Jean-Paul Sartre, Frantz Fanon, Michel Foucault, Herbert Marcuse, Noam Chomsky, and many contemporary Greens). Herman writes:

> The historical pessimist sees civilization's virtues under attack from malign and destructive forces that it cannot overcome; cultural pessimism claims that those forces form the civilizing process from the start. The historical pessimist worries that his own society is about to destroy itself, the cultural pessimist concludes that it needs to be destroyed.

Thus spake Nietzsche: "There is an element of decay in everything that characterizes modern man." And: "Are we not straying as through an infinite nothing? Do we not feel the breath of empty space? Has it not become colder? Is not night continually closing in on us?"

A standard eco-pessimist could announce almost triumphantly in 1992:

> Modern humanity is rapidly destroying the natural world on which it depends for its survival. Everywhere on our planet, the picture is the same. Forests are being cut down, wetlands drained, coral reefs grubbed up, agricultural lands eroded, salinized, desertified, or simply paved over. Pollution is now generalized—our groundwater, streams, rivers, estuaries, seas and oceans, the air we breathe, the food we eat, are all affected.

Just about every living creature on earth now contains in its body traces of agricultural and industrial chemicals—many of which are known or suspected carcinogens or mutagens.

As a result of our activities, it is probable that thousands of species are being made extinct every day. Only a fraction of these are known to science. . . . By destroying the natural world in this way we are making our planet progressively less habitable. If current trends persist, in no more than a few decades it will cease to be capable of supporting complex forms of life.

(That was Edward Goldsmith's opening salvo in *The Way: An Ecological World-view*. His worries are accurate individually, but they are selective, one-sided, and overaggregated into a paralyzing spasm of angst.)

● Arthur Herman traces the origin of romanticism and its decay narrative to one man and one event—Jean-Jacques Rousseau and the French Revolution of 1789. Rousseau embraced an imaginary primitivism and declared, "Everything degenerates in the hands of men." His vision of a return to innocence and freedom seemed to be at hand with the overthrow of the French monarchy. The intelligentsia of Europe thrilled to the coming of a new dawn in 1789, and then watched it turn into blood and terror by 1793. With that trauma, the romantic stance became one of despair and defiance, and it has remained so ever since.

Following the deep seam of romanticism through successive centuries, Herman finds it leading through Oswald Spengler's *Decline of the West* (1918) directly to Nazi Germany. "Hitler's generation was the first European generation raised on cultural pessimism."

There is a troubling Green thread in the Nazi movement. I first came across it in 1977 with an article I ran in *CoEvolution* on the German *wandervögel* (wanderbirds)—young hippielike back-to-the-land romantic strivers of the late nineteenth century who were all too easily co-opted into the Hitler Youth. I learned from Herman's book that biologist Ernst Haeckel, coiner of the word *ecology* (*oekologie*, 1866), championed eugenics and selective euthanasia to purge an imperiled Europe of "degenerates such as Jews and Negroes." According to Peter Coates in *Nature: Western Attitudes Since Ancient Times* (2004), "Nazi Germany led Europe in the creation of nature reserves and the implementation of progressive forestry sensitive to what we would now call biodiversity."

Summarizing the ongoing debate about "the Green face of Nazism," Nils Gilman at Global Business Network wrote me that,

> The key and I think undebated points are these: (1) the Nazis used their green credentials to win and widen their popular support, (2) virtually no one at the time, inside or outside Germany, saw any contradiction between the Nazis' environmentalism and the rest of their political program. In sum: while there's obviously no necessary connection between eco-friendliness and fascism/nativism, there are lots of ways in which the two movements can and have connected historically, and may again in the future.

• How times change. Germany is once again the Greenest country in Europe, but this time the political framework is so leftist that the powerful Green party members, *Die Grünen*, are commonly called watermelons: green on the outside, red on the inside. That flip is common in the world. In the old days, conservation was conservative, the proper activity of duck hunters and Teddy Roosevelts. And progress used to belong to progressives; but then it frightened them, and they turned on it. They came to oppose what they viewed as the technological threats of progress, the despoliation of nature by progress, and the capitalist engine of progress. That in turn offended the conservatives, who were fond of capitalism, and opposing the newly antiprogress progressives meant opposing their environmental programs as well. The flip was complete.

It has become a problem. Worldwide, the political stereotype these days is that Green equals left, left equals Green, and right equals anti-Green. That may be helpful for liberals, grounding them in the science and practice of natural systems, but it blinds conservatives and badly ham-

pers Green perspective. Becoming politically narrow limits Greens' thinking and marginalizes their effectiveness, because whatever they say is
automatically dismissed by anyone who has doubts about liberals. Countless conservatives refused to take climate change seriously because they
couldn't abide the idea of Al Gore being right.

I saw a version of this narrowness played out after 1966, when I was
inspired by a rooftop LSD trip to distribute buttons that read, "Why haven't
we seen a photograph of the whole Earth yet?" Everyone in the New Left
opposed Kennedy's space program, seeing it (correctly) as a cold war episode that they thought (incorrectly) was being carried out to no good purpose by crew-cut military squares. (Only Abbie Hoffman disagreed with
his compatriots: "Are you kidding? We're going to the fucking MOON!")
Environmentalists joined the leftist opposition to the space program: "We
have to clean up the Earth before we can leave it."

The exception was Jacques Cousteau, the pioneer of underwater exploration. In a 1976 interview for *CoEvolution*, he told me that in the 1960s his
fellow ocean specialists were scandalized by the expense and irrelevance
of the U.S. space program, but he supported it for philosophical reasons
that quickly became practical. Cousteau realized that satellites were the
only way to monitor the health of the oceans.

Despite their best efforts to shut it down or ignore it, environmentalists gained more from the space program than anyone else, and
sooner. Directly inspired by the 1969 photos of Earth from space, the
first Earth Day in 1970 attracted 20 million Americans to the rallies,
and the environmental movement took off, with a planetary icon and
a coherence it has maintained ever since. Robert Poole wrote in *Earthrise* (2008): "As soon as the Earth became visible . . . it began to acquire
friends, starting in 1969 with Friends of the Earth. The years 1969–72
saw no fewer than seven major national environmental organizations
come into being."

What made Cousteau prescient about what the perspective from space
would bring? He had no allergy to new technology: He was the inventor of
the scuba gear that made underwater exploration possible. His explorer's
heart saw space as the next ocean, and his scientific perspective made him
ask what satellites could do for him. Being apolitical, he was free of loyal-

ties to any narrow agenda. To disagree with his scientific peers was not a violation of solidarity but part of his job as a scientist.

Solidarity is a leftover idea of the left—"Which Side Are You On?" was a union song—that has no place in the environmental movement. It led Friends of the Earth in Britain to throw away their trustee Hugh Montefiore over nuclear power. (He supported nuclear to head off climate change.) The man who fired Montefiore, FOE director Tony Juniper, said that debate was welcome within the organization but not in public. That strikes me as a self-defeating practice. It is more important for an organization or a movement to be right than to be consistent, and figuring out what is right takes debate, as open as possible, because what is right keeps on changing as circumstances change.

• A romantic stance, or a political agenda, is fine for giving people a sense of identity and motivating their efforts; but it's poor at solving problems. "One of the points of pragmatism is that there is no escape from the need to wrestle seriously with the particulars of a given problem," writes Daniel Farber in his law book *Eco-pragmatism* (2000).

Paul Hawken has one of the great business-card stories:

> I have given nearly one thousand talks about the environment in the past fifteen years, and after every speech a smaller crowd gathered to talk, ask questions, and exchange business cards. The people offering their cards were working on the most salient issues of our day: climate change, poverty, deforestation, peace, water, hunger, conservation, human rights, and more. They were from the nonprofit and nongovernmental world, also known as civil society. They looked after rivers and bays, educated consumers about sustainable agriculture, retrofitted houses with solar panels, lobbied state legislatures about pollution, fought against corporate-weighted trade policies, worked to green inner cities, or taught children about the environment. Quite simply, they were trying to safeguard nature and ensure justice.

Hawken kept the growing pile of cards until they provoked him into action. As he researched his book *Blessed Unrest* and set about building an online database of such organizations, he began to realize that there are over a million of them loose in the world. They are flourishing because of their specificity, because they wrestle with particulars. They are invisible for the same reason, and effective for the same reason. As Hawken notes, "Feedback loops are short, learning is accelerated." Unfettered by ideology, slogans, fame, or even an aggregate name, the organizations live by improvisation and focus on results. Their story is about improvement, not decline.

✦

Science is the only news. When you scan a news portal or magazine, all the human interest stuff is the same old he-said-she-said, the politics and economics the same sorry cyclical dramas, the fashions a pathetic illusion of newness; even the technology is predictable if you know the science. Human nature doesn't change much; science does, and the change accrues, altering the world irreversibly.

In stark contrast to romantic cultural pessimism, science is imbued with a double optimism. One part is the scientific process itself, driven by accelerating capability: science makes science go faster and better. The other part is the content—much of what is discovered is either good news or news that can be made good, thanks to ever-deepening knowledge, tools, and techniques. Because the findings of science are not just matters of opinion, they sweep past systems of thought based only on opinion. The swarming edges of science pose ever more and better questions, better put. They're phrased to elicit hard answers, the answers get found, and the questioners move on.

No wonder static, self-obsessed romanticism acts so threatened by science. It is. A romantic loves the tree, not its genome. A scientist loves both.

Literary agent John Brockman points out another angle on the news from science:

> Through science we create technology and in using our new
> tools we recreate ourselves. But until very recently in our his-

tory, no democratic populace, no legislative body, ever indicated by choice, by vote, how this process should play out. Nobody ever voted for printing. Nobody ever voted for electricity. Nobody ever voted for radio, the telephone, the automobile, the airplane, television. Nobody ever voted for space travel. Nobody ever voted for nuclear power, the personal computer, the Internet, email, the Web, Google, cloning, the sequencing of the entire human genome.

Science proposes, society disposes.

● Environmentalists do best when they follow where science leads, as they did with climate change. They do worst when they get nervous about where science leads, as they did with genetic engineering. You can see the romantic affliction at work right there. Climate change fit in with the romantic idea of decline and disaster. Genetic engineering looked like Dr. Frankenstein's sin against nature in Mary Shelley's classic romantic story.

I would like to see the environmental movement—and indeed everybody—become fearless about following science. Part of that process lies in learning which scientists and which research to track most closely.

Our first duty is to be wary of confirmation bias—the inclination to notice and believe whatever supports our current theory, and ignore or disbelieve everything that doesn't support our views. It takes harsh self-discipline to overcome. "Darwin writes in his autobiography," reports Bell Labs researcher Richard Hamming, "that he found it necessary to write down every piece of evidence which appeared to contradict his beliefs because otherwise they would disappear from his mind."

Another hard task is to beware of the plausible little stories we tell ourselves. In 1998 I was sure that the Y2K bug would be a major problem, and said so in public and to clients of Global Business Network. My wrong prediction was based on a neat little story I told myself. My own PC was pathetically vulnerable to bugs in the software, which could lead to a cascade of problems ending in the blue screen of death. Surely, I presumed, the huge old mainframes of the world and their ancient software would be even more vulnerable to a bug as deeply embedded as I thought Y2K must be. The world was facing the blue screen of death! I should have listened to Danny Hillis, who has designed whole computer platforms. He predicted that Y2K would lead, at worst, to some dog licenses not being renewed on time. I should have listened at a dinner in 1998 with senior Amazon.com engineers. One of them responded to my rant about Y2K by sweetly inquiring, "Do you also believe in fairies?"

Lesson: Question convenient fables; listen most closely to the scientists who know the facts best, have studied them longest, and aren't biased by an agenda or an employer with an agenda.

Following the publication of his *Plows, Plagues and Petroleum*, climatologist William Ruddiman found he was suddenly the target of a barrage of propaganda:

> These newsletters opened a window on a different side of science, a parallel universe of which I had been only partly aware. The content of these newsletters purports to be scientific but actually has more in common with hardball politics.
>
> Most of these articles come from contrarian web sites that receive large amounts of financial support from industry sources. In many cases, the authors are paid directly by industry for the articles they write. . . .
>
> This alternative universe is really quite amazing. In it, you can "learn" that CO_2 does not cause any climatic warming at all. You can find out that the world has not become warmer in the last century, or that any warming that has occurred results from the Sun having grown stronger, and not from rising levels of greenhouse gases. One way or another, most of the basic findings of mainstream science are rejected or ignored.

Quasi-scientific propaganda against climate change is no different from quasi-scientific propaganda against genetic engineering. Both try to harness science to a political agenda.

Eliminating "bought" scientists still leaves plenty of legitimate scientists who disagree, sometimes fiercely, on any particular issue. "In science,"

John Brockman reminds those confused by the combat, "debate is the way people work together, the way they advance their ideas." Geneticist Pamela Ronald, in *Tomorrow's Table*, offers a short course on how to distinguish science from rumor and how to weigh a scientific debate. Her major points:

> Examine the primary source of information. . . . Ask if the work was published in a peer-reviewed journal. . . . Check if the journal has a good reputation for scientific research. Determine if there is an independent confirmation by another published study. . . . Assess whether a potential conflict of interest exists. . . .

I would add to that: Watch for trends. Over time, what does the growing preponderance of evidence indicate? Is a consensus among scientists emerging?

• Environmentalists were right to be inspired by marine biologist Rachel Carson's book on pesticides, *Silent Spring*, but wrong to place DDT in the category of Absolute Evil (which she did not). Most of her scientific assessments proved right, some didn't—such as her view that DDT causes cancer. In an excess of zeal that Carson did not live to moderate, DDT was banned worldwide, and malaria took off in Africa. Quoted in a 2007 *National Geographic* article, Robert Gwadz of the National Institutes of Health said, "The ban on DDT may have killed 20 million children." These days, environmental organizations such as World Wildlife Fund support the judicious antimalaria use of DDT on household walls as one element of "integrated vector management," along with bed nets, larvicides in standing water, and other measures that could lead to totally eradicating the disease from the world. When malaria disappears, so can DDT.

Science too often gets perverted by politics. You can see it in two exemplary case studies: *Lament for an Ocean: The Collapse of the Atlantic Cod Fishery: A True Crime Story* (1998), by Michael Harris, and *Degrees of Disaster: Prince William Sound: How Nature Reels and Rebounds* (1994), by Jeff Wheelwright. In the cod story, you learn that while independent scientists were predicting the collapse of the declining fishery in the 1980s, Canadian politicians and government scientists pretended all was well, in part because they felt a responsibility to protect the jobs of the fishing communities of Newfoundland. The collapse came in 1989. Fishing for cod was totally banned in 1992, but the cod fishery still has not recovered, and it may never; tens of thousands of jobs were lost permanently in Newfoundland and other Maritime Provinces.

A similar sequence is playing out in one fishery after another—haddock, tuna, salmon, rockfish. Three fishery-preserving strategies that show promise are: ocean reserves that ban all fishing in designated areas and allow stocks to recover; a system of catch shares called "individual transferable quotas," which has already saved the halibut fishery in Alaska; and carefully managed mariculture. As Jacques Cousteau told me in 1976, "Fishing is hunting. . . . It must be eliminated completely and replaced by farming if we are to be civilized. What we call civilization originated in farming. We are still barbarians in the sea."

Degrees of Disaster, the close-up story of the *Exxon Valdez* oil spill, is replete with awkward truths that didn't make it into the warring scientific reports so sumptuously funded by both sides of the controversy. The massive cleanup efforts did more environmental harm than the spill itself, though they did provide an economic boom for the Prince William Sound region. The biologically richest ocean habitat in the area was *inside* the emptied cargo holds of the grounded ship: an entire food chain from bacteria up to herring and salmon was feeding on the oil. People were worried about aromatic hydrocarbons in wild salmon after the spill, but it turned out that the highest level reached was one ten-thousandth of what was normally found in the local traditionally smoked salmon. The real lesson of the oil spill at Prince William Sound is how resilient many natural systems are and how rapidly they bounce back when human pressure backs off even a little.

Environmentalists do a public service when they help to depoliticize science. In 2008, Britain's Labour government was poised to ban plastic grocery bags because they were thought to get into the sea and entangle marine birds and mammals. A marine biologist at Greenpeace, David Santillo, spoke up: "It's very unlikely that many animals are killed by plastic bags. On a global basis, plastic bags aren't an issue." The government action, it turned out, was based on a misreading of a Canadian report that 100,000 marine animals were killed by entanglement in discarded fishing gear and nets in a period of four years. Grocery bags had nothing to do with the problem.

Scientists freely criticize each other and lambaste anti-Darwinians, but they are weirdly polite with environmentalists. It smells of condescension. Every biologist I know is dismayed by the Green campaign against genetic engineering, but the only one who speaks out is Peter Raven. Climatologists see the need for nuclear power, but the only ones who publicly criticize environmentalists for their opposition are Jim Lovelock and James Hansen. It's time to stop coddling environmentalists. Their motivation is not fragile. Their effectiveness will increase to the degree that they are armed with scientific sophistication and discipline. If they are treated as peers by scientists—which means harshly—they might become peers.

◆

Years ago, environmentalists hated cars and always tried to ban them. Then Amory Lovins came along. He decided that the automobile was the perfect leverage point for large-scale energy conservation, and he set about designing and promoting radically more efficient cars. Lovins single-handedly converted the environmental movement from loathing the auto industry to fruitfully engaging with it. That's the engineering approach. Instead of yelling "Stop!" engineers figure out what the problem is, and then make it go away. They don't have to argue about what is wrong; they show what is right.

Another Green issue that responded well to an engineering approach is the design of buildings. Paul Hawken describes how the LEED rating system came into existence:

> Buildings . . . use 40 percent of all material and 48 percent of
> the energy in the United States alone. In 1993 David Gottfried,
> a successful but disillusioned developer, and Rick Fedrizzi, an
> executive at the Carrier corporation, gathered a small group of

architects, suppliers, builders, and designers in order to create a rigorous set of green building standards. Today, the US Green Building Council (USGBC) comprises 6,200 institutional members and 85,000 active participants, and green building councils exist in Japan, Spain, Canada, India, and Mexico. No one has done the metrics, but in its short life USGBC may have had a greater impact than any other single organization in the world on materials saved, toxins eliminated, greenhouse gases avoided, and human health enhanced. It collaborates with designers, architects, and businesses—not always easy, because their movement means the company's products must change—in order to define and incrementally raise the environmental standards of green buildings by means of a rating system called Leadership in Energy and Environmental Design (LEED).

Now developers and city planners compete to earn the highest LEED ratings—Silver, Gold, or Platinum. The bragging rights are worth the extra money they spend.

● For many environmentalists, the entry to an engineer's way of thinking came with a book by Janine Benyus, *Biomimicry: Innovation Inspired by Nature* (1997). The book is meant to inspire engineers to study nature for design ideas, but along the way, it teaches nonengineers to respect engineering. "Unlike the Industrial Revolution," Benyus writes, "the Biomimicry Revolution introduces an era based not on what we can *extract* from nature, but on what we can *learn* from her." Among the examples she explores are

solar cells copied from leaves, steely fibers woven spider-style, shatterproof ceramics drawn from mother-of-pearl, cancer cures compliments of chimpanzees, perennial grains inspired by tallgrass, computers that signal like cells, and a closed-loop economy that takes its lessons from redwoods, coral reefs, and oak-hickory forests.

And she extracts nine basic principles:

> Nature runs on sunlight. Nature uses only the energy it needs. Nature fits form to function. Nature recycles everything. Nature rewards cooperation. Nature banks on diversity. Nature demands local expertise. Nature curbs excesses from within. Nature taps the power of limits.

(In this formulation, Nature looks suspiciously like a liberal Democrat. Since Benyus got her degree in forestry, she could probably compile a fine Devil's Dictionary version of her list. Here's mine:

> Nature rewards efficiency: The most efficient way of life is that of a parasite. Nature is merciless: The best way to control your niche is to annihilate competitors. Nature honors property: Stake out your turf and defend it with your life, or starve. Nature favors opportunistic invaders, such as kudzu and humans. Nature is parsimonious: Eat excess children.

Through the success of the book and a campaign of lectures and workshops by Benyus and coworkers, biomimicry established itself firmly and effectively. A 2008 book from the National Academy of Sciences noted:

> In the last decade, there has been an explosion of information about unusual natural structures that are superstrong, superadhesive, superhydrophobic, superhydrophilic, super-efficient, self-cleaning, self-healing, and self-replicating, with superior designs and intricate shapes. Biological materials are also often multifunctional, a characteristic highly desirable in artificial materials and processes.

Along the way, researchers discovered that nature is extremely difficult to mimic in detail. You can fruitfully steal ideas and mechanisms, but natural processes defy simple imitation because they are the irrational product of timeless evolution rather than design. A 2008 article in *National Geographic* spelled out the problem:

> The main reason biomimetics hasn't yet come of age is that from an engineering standpoint, nature is famously, fabulously, wantonly complex. . . . To make the abalone's shell so hard, 15 different proteins perform a carefully choreographed dance that several teams of top scientists have yet to comprehend. The power of spider silk lies not just in the cocktail of proteins that it is composed of, but in the mysteries of the creature's spinnerets, where 600 spinning nozzles weave seven different kinds of silk into highly resilient configurations.

Birds showed us that heavier-than-air flight is possible, and their structure suggested two wings at the front of a flying machine and a stabilizer in back. But the bird's wing flap was too hard to imitate, so for propulsion humans devised rotating wings (the propeller)—something new to nature. (The idea presumably came from the previous invention of windmills. Instead of taking energy out, put energy in; run the vanes backward, and make wind.)

● As the designers and builders of our fabricated infrastructure, engineers are comfortable with taking on natural infrastructure, nudging nature's processes where needed with human ingenuity. Got a flood-prone river? Moderate it with dams and collect some electricity, free of greenhouse gases, at the same time.

Greens interested in engineering solutions for environmental problems but nervous about their hubris will watch with fascination and horror what happens over the coming decades in China, a nation run by engineers rather than lawyers. *New Scientist* reported in 2007:

> Until this year's 17th National Congress of the Communist Party of China . . . every member of the central bastion of power— the Standing Committee of the Politburo—was an engineer by

training. President Hu Jintao is a graduate of Beijing's Tsinghua University, often referred to as China's MIT, while the premier, Wen Jiabao, trained as a geologist.

When China goes Green, it goes Green big. "The Chinese have purchased 35 million solar water heaters, more than the rest of the world combined," said a 2007 article in *Seed* magazine, and

> China currently ranks sixth in the world in total wind power production, . . . but by 2020 it aims to increase its share by 1,200 percent, to 30,000 megawatts of power. . . . In 2000, [Beijing] took 26,000 heavily polluting minibuses off the road in a week. . . . Across China, the government is constructing massive solar-and biofuel-powered eco-cities 30 times the size of the largest green communities elsewhere in the world.

When Kevin Kelly was traveling in China in 2006, he found that every elementary school in every village had a sign over the door in Mandarin with the following guidance:

LOOK UP TO SCIENCE.
CARE FOR YOUR FAMILY.
RESPECT LIFE.
RESIST CULTY RELIGION.

(That raises the question of whether environmentalism is a culty religion. It clearly won't be in China.)

● Meanwhile, in North America and Europe, environmental problems are treated as commercial opportunities, and Green entrepreneurs lead the way. Engineers are being hired in droves. They don't know or care much about environmental traditions, causes, or romantic attitudes. Because they are interested in solving problems, not in changing behavior, technology is the first thing they reach for when looking for a solution.

Green venture capitalist in Silicon Valley, Vinod Khosla, told *ntist* that improvements in energy efficiency or changes in the duce only small, incremental gains. "A new technology, on the other hand, can make a 200 per cent or a 400 per cent or a 1000 per cent difference."

A notable example of rethinking a whole industrial domain in environmental terms is the discipline of "Green chemistry," named and defined by Paul Anastas in 1991, when he was head of industrial chemistry at the Environmental Protection Agency. His now-canonical "Twelve Principles of Green Chemistry" are not known to most environmentalists; but they should be, because they outline exactly how to head off the worst form of pollution after greenhouse gases, namely toxic chemicals. ("Minimize waste by using catalytic reactions," advises one principle. "Maximize atom economy," says another.)

When environmentalists are wrong, it is frequently technology that they are wrong about, and they wind up supporting parochial Green goals at the cost of comprehensive ones. That happened with space technology, nuclear technology, and genetic engineering. If your default position on a new technology is suspicion, you forfeit the ability to deploy it for your own purposes. "The environmental movement has so far concentrated its attention upon the evils that technology has done rather than upon the good that technology has failed to do," says Freeman Dyson. But focusing on Green technological opportunity requires a shift in attitude toward novelty.

What a joy it would be to see a new generation of hardcore environmentalists as intrigued by new infrastructure-scale technologies as they are (and as I am) by new ultralight hiking gear made of titanium and Pertex. Their seriousness would be driven not by the romantic love of decline but by a desire to grab progress and direct it toward Green goals. Their potential tools would range from the molecular (genes and atoms) to the cosmic (solar winds and dark energy). They would follow their cellphones into the core of life.

◆

"How you think matters more than what you think," says political scientist Philip Tetlock. The most important distinction in quality of judgment, he declares, was first expressed by the ancient Greek poet Archilochus: "The fox knows many things; the hedgehog one great thing." Hedgehogs have a grand theory they are happy to extend into many domains, relishing its parsimony, expressing their views with great confidence. Foxes, on the other hand, are skeptical about grand theories, diffident in their forecasts, and ready to adjust their ideas based on actual events. Hedgehogs don't notice or care when they're wrong. Foxes learn. Hedgehogs are great proponents, but foxes are invariably better forecasters and policy makers.

The authoritative book on this subject is Tetlock's *Expert Political Judgment* (2005). From his perspective as a psychology researcher, Tetlock watched political advisers on the left and the right make bizarre rationalizations about their wrong predictions at the time of the rise of Gorbachev in the 1980s and the eventual collapse of the Soviet Union. (Liberals were sure that Reagan was a dangerous idiot; conservatives were sure that the USSR was permanent.) The whole exercise struck Tetlock as what used to be called an "outcome-irrelevant learning structure." No feedback, no correction.

So Tetlock took advantage of getting tenure at the University of California–Berkeley to start a long-term research project, now twenty years old, to examine in detail the outcomes of expert political forecasts about international affairs. He studied the aggregate accuracy of 284 experts making 28,000 forecasts, looking for pattern in their comparative success rates. Most of his findings were negative—conservatives did no better or worse than liberals; optimists did no better or worse than pessimists. Only one pattern emerged consistently.

Tetlock borrowed the hedgehog-fox distinction from essayist Isaiah Berlin, who gave as examples of single-minded hedgehogs Plato, Dante, Hegel, and Proust; as open-minded foxes, Aristotle, Shakespeare, Voltaire, and Joyce. The aggregate prediction success rate of foxes is significantly greater, Tetlock found, especially in short-term forecasts. Hedgehog experts are not only worse prognosticators than fox experts (especially in long-term forecasts); *they even fare worse than normal attention-paying dilettantes* like you and me—apparently blinded by their extensive expertise and beautiful theory. Furthermore, foxes win not only in the accuracy of their predictions but also the accuracy of the likelihood they assign to their predictions; in this they exhibit something close to the admirable discipline of weather forecasters.

The value of hedgehogs is that they occasionally get right the farthest-out predictions, but that comes at the cost of a great many wrong far-out predictions. The charismatic expert who exudes confidence and has a great story to tell is probably wrong about what's going to happen. The boring expert who afflicts you with a cloud of howevers is probably right. "There is an inverse relationship between what makes people attractive as public presenters and what makes them accurate in these forecasting exercises," Tetlock told a San Francisco audience. He added that hedgehogs annoy only their political opposition, while foxes annoy across the political spectrum.

• As with political experts, so with environmental experts. Two celebrated commentators worth analyzing are Bjørn Lomborg, author of *The Skeptical Environmentalist* (2001), and Amory Lovins. I once heard a Lomborg fan ask Jared Diamond what he thought of Lomborg's book. Diamond's answer, as I recall it, was: "The problem is, Lomborg argues from details. He says the ecological collapse of Easter Island offers no general lessons because it was due to the fragility of one kind of palm tree. Arguing with that kind of reasoning is like arguing with a Creationist about some inverted geology they've found in Texas that they say disproves Darwin. If you take the time to research their example and disprove their interpretation, you find out it doesn't matter. They don't care. They've found some other detail they think supports their theory." (Diamond went on to praise Lomborg for his efforts to focus the world's attention on eradicating malaria in Africa.)

Lomborg's constant message, delivered with sweet reasonableness, is that environmentalists mean well, but they always exaggerate dangers. With his background in statistics, he drills happily into data, making his case with a profusion of details. That kind of expertise can make hedge-hogs overconfident, Tetlock writes:

> They have so much case-specific knowledge at their finger-tips, and they are so skilled at marshalling that knowledge to construct compelling cause-effect scenarios, that they talk themselves into assigning extreme probabilities that stray fur-ther from the objective base-rate probabilities. As expertise rises, we should therefore expect confidence in forecasts to rise faster, far faster, than forecast accuracy.

One scientist who has taken the trouble to refute Lomborg in his own mode is water conservationist Peter Gleick in a lengthy 2001 review of *The Skeptical Environmentalist* for the Union of Concerned Scientists. Gleick meticulously dissects the book's "selective use of data, misuse of data, mis-interpretations, inappropriate precision, errors of fact."

As for Amory Lovins, he doesn't just argue from details, he backs up a truck full of numbers and citations and dumps them on you, saying that if you won't master them, you can't possibly argue with him. Events have proven him profoundly right about energy efficiency and conservation and wrong in his forecasts about nuclear power. I predict that he will maintain a hedgehog stance on that subject and never have a good word to say about nuclear (nor, I expect, will Lomborg ever have a good word to say about environmentalists). If I have the pleasure of being wrong about Lovins, I'll bet his change of opinion develops around microreactors, which fit in with his views about distributed micropower.

Tetlock writes that hedgehogs deploy a routine set of excuses when proven wrong: "I was almost right"; "I was just off on timing"; "I made the right mistake" (right policy, wrong prediction); "Happenstance went against me." Each excuse provides an opportunity to explain one more time the deep rightness of the original theory.

● Scientists are trained to be foxes. One outspoken voice on climate is Stanford climatologist Stephen Schneider. In 1971 he wrote an influential paper that predicted global *cooling*, based on the previous three decades of cold weather and his model of how the increase of dust and particles in the air (called aerosols) from human activity might trigger an eventual ice age. In 1974 he publicly retracted the paper, having become convinced that his model overestimated aerosol effects and underestimated carbon dioxide effects. With better data and a better model, he reversed his position to extreme concern about global warming, which he maintains to this day.

Naturally, climate denialists mocked him about being so flexible (for political reasons, they assumed), just as an English legislator once chided John Maynard Keynes for reversing his position on money policy during the Great Depression. Keynes replied, "When the facts change, I change my mind. What do you do, sir?"

"Whenever I start to feel certain I am right . . . a little voice inside tells me to start worrying," one of Phil Tetlock's respondents told him—a statement he considers "a defining marker of the fox temperament." Another comes from the French fox Voltaire: "Doubt is not a pleasant condition, but certainty is absurd."

Every interview with a public figure should include the question "What have you been wrong about, and how did that change your views?" The answer will tell us if the person is intellectually honest or a tale spinner with delusions of infallibility. Let me quickly furnish a partial list of things I've been wrong about in public. In the 1960s, I pushed communes as a path to the future, Buckminster Fuller domes as habitable, and cocaine as harmless. In the 1970s, I was sure the 1973 oil crisis would lead to police in the streets of the United States, that nuclear power was bad, and that small was always beautiful, villages especially. I was totally wrong about the Y2K bug in 2000. In 2003 I was so sure that a Democrat would win the 2004 presidential election that I made a public bet about it. Hey, I was just off on timing.

Fessing up aids learning. From these mistakes and others, I have learned to suspect my excesses of optimism and pessimism. Apparently I often think that societies catch on faster than they do, and that large complex systems are more brittle than they are. Bear in mind I might be wrong that way about climate. And many of my faulty opinions turn out to be based on ignorance; dismissing nuclear was one of those.

● One source of confusion for people is that the views of hedgehogs are strongly stated and strongly held, while the views of foxes are modestly stated and loosely held. Guess who gets audience share. What we need is more brazen foxes who don't mind strongly stating their loosely held views (this book tries to be an example), and audiences that honor honest opinion change. When some pontificator begins, "As I've always said, . . ." the right response is "Uh oh."

Failure to acknowledge a mistake is paralyzing. During the Iraq War, a friend who consulted for the George W. Bush White House told me, "The Neoconservatives don't even try to say they were right about Iraq anymore. They spend all their effort trying to prove they weren't wrong. That means you never change policy because you can never have the discussion that begins, 'Well, Plan A didn't work. What have we got for Plan B?' " It's a bad idea to appoint or elect hedgehogs to power positions.

The most powerful fox I've known personally was California governor Jerry Brown. He had a remarkable technique with protesters, based on sheer curiosity. Whenever he saw a lineup of demonstrators, he would walk over and engage them: "Tell me what you're concerned about." Someone would launch into their rant, and Brown would listen. After a bit he would interrupt, "Let me see if I got it. You're saying that . . ." And he would state their position, often with greater clarity and eloquence than theirs. They would just melt. They'd been heard! He got it! They knew he probably wouldn't change his position on the issue they were protesting, but, who knows, he might, because Governor Brown was famous for occasionally reversing his opinion and his policy, in response to events.

I used a variant of Brown's approach in designing the debate format for the Seminars About Long-term Thinking I run for Long Now in San Francisco. We had one such debate on the Greening of nuclear power and another on synthetic biology. Whichever debater goes first holds forth for fifteen minutes and then is interviewed for ten minutes by the second debater, who has to conclude by summarizing the first debater's argument *to the first debater's satisfaction:* "You got it." Then they reverse roles.

Audiences love it. They relish watching public figures struggle to state an opposing opinion right out loud, without sarcasm. Better still, the shared probe for depth of understanding of the issue replaces the usual win-lose mutual deafness of public debates. As a result, audience members find that the hard edges of their own opinions start to soften.

There is nowhere a good venue for honest debate about environmental issues. News media like loud fights between glib hedgehogs, not polite debate that reaches for depth. Environmentalist organizations don't have enough money to host big conferences. Maybe some Green philanthropists could help with that. The most direct solution might be for scientific conferences, which are well funded, to invite more environmentalists to come and debate formally in their venue.

✦

If Greens don't embrace science and technology and jump ahead to a leading role in both, they may follow the Reds into oblivion. They need to become early adopters of new tools and adventurous explorers of new situations. Instead of always saying "No" and "Stop," their strategy can be to affirm and redirect. They could give a new technology the benefit of the doubt—but never throw away their doubt—using it to shape the technology in gentler ways to better ends.

Rather than cherishing the role of romantic rebels and avoiding government, Green activists should leap into government, seeking to emulate the all-embracing government-run Green plans of New Zealand and the Netherlands. They can take inspiration from governmental foxes like Franklin Roosevelt, described thus by his contemporary Isaiah Berlin:

> Roosevelt stands out principally by his astonishing appetite for life and by his apparently complete freedom from fear of the future; as a man who welcomed the future eagerly as such, and conveyed the feeling that whatever the times might bring, all would be grist to his mill, nothing would be too formidable or

crushing to be subdued and used and moulded into the pattern of the new and unpredictable forms of life into the building of which he, Roosevelt, and his allies and devoted subordinates would throw themselves with unheard-of energy and gusto.

It wasn't just attitude. He made it work.

Footnotes for this chapter may be found at **www.sbnotes.com**.

·8·
It's All Gardening

Never pick your herbs from the first bush you come to.

—Thomas Banyacya, Hopi

Wilderness can't take care of itself. It has to pay its way.

—ecologist Daniel Janzen

Ecosystem engineering is an ancient art, practiced and malpracticed by every human society since the mastery of fire. We would be fools if we repeat their mistakes and just as foolish if we ignore some of the brilliant practices that worked for them. But first we have to figure out what really happened, and that requires discarding some cute stories we like to tell ourselves about indigenous people.

Romanticism has been particularly unkind to American Indians. Most people get their ideas about Native American life from books like *The Teachings of Don Juan* (1968), *The Education of Little Tree* (1976), *The Tracker* (1978), *Medicine Woman* (1981), *Buffalo Woman Comes Singing* (1991), *Spirit Song* (1985), and *Seven Arrows* (1972). In the same category, but concerning Australian Aborigines, is *Mutant Message Down Under* (1994). These books sold in the millions; two were recommended by Oprah. They're all bogus. The nearly universal tale is that of a white seeker taken under the wing of a native spiritual teacher who imparts ancient wisdom, and the reader is let in on the secret knowledge of the Yaqui, Cherokee, Apache, Cree, Crow, Chippewa, Cheyenne, or Australian aborigines. Each book has been condemned by the tribe in question as insulting fiction.

What is it the readers are longing for, and why do we settle so easily for lies? (I confess to being taken in by Carlos Castaneda's first Don Juan book; his sequels were increasingly transparent New Age drivel.) I think we seek the wrong connection in these books. Esoteric knowledge, even when it's genuine, is of little use outside the culture it's rooted in, and to seek it from outside is a trespass. What is not in the books and is in the tribes is their *groundedness*. They know how to live where they live. That's the lore worth learning. Many assume that because the Indian "wisdom" they see in popular books and movies is clearly phony, any real traditional knowledge is gone from the modern world. I have found otherwise.

• In 1963, after I left the army, I did some photography on assignment at the Warm Springs Reservation in Oregon. The contemporary Indian reality I saw there was a revelation to me. So in the years following, I spent all my summers with various native communities—Wasco, Paiute, Navajo, Hopi, Zuni, Taos, Jicarilla Apache, Papago (now Tohono O'odham), Ute, Blackfoot, Sioux, Cherokee, Ponca. I joined the Native American Church (which uses peyote as a sacrament), and I married an Ottawa Indian mathematician with raven hair, Lois Jennings. In the winters of those years, I performed a multimedia show called *America Needs Indians* in museums and nightclubs.

What was all that about? Playing Indian is not really an option when you're with real ones; I guess I was learning how to be American in a way that had nothing to do with the Pledge of Allegiance. One freezing December night at Zuni Pueblo in New Mexico, I saw ten-foot-high gods, the Shalakos, being welcomed into homes especially rebuilt to give them room to dance, while Mudhead clowns with assholes for eyes mocked everybody. The sacredness of the ceremony gave the comics their bite, and that bite paradoxically made the gods realer. I was shivering from more than cold.

Most North American tribes are matrilineal. I learned something about that at a peyote meeting. The vice president of the Native American Church in the 1960s was a Navajo named Hola Tso. When I showed up at his home, he extended the hospitality I found everywhere in Indian country and invited me to a meeting.

The format of the peyote meeting is a work of genius. Universal throughout North America, it is the opposite of esoteric knowledge. The officers of the meeting—roadman, cedarman, drummer, and fireman— do not have powers; they are administrators strictly, though with great art. Only one officer represents something larger, and that is held back till the end of the meeting. It's a long night, progressing formally through midnight water call and the emotional crisis at three A.M. and the hard-won ascent toward dawn. As sunrise brightens the east-facing doorway of the hogan or tipi, a woman who is also Peyote Woman comes in, bearing food.

At Hola Tso's meeting, the authority in her voice floored me. The fragrance of fresh fruit and fresh-cooked meat penetrated our heightened senses, but she wasn't going to let us have any until we had heard what she had to say. She reminded us where the food came from (her). She reminded us where life comes from (her). She waited while that sank in. Then she passed the food. My young male mind was liberated by Peyote Woman.

All I got, of course, was tiny glimpses. It's all still there: the banality of the "rez" that Sherman Alexie writes about, pan-Indian powwows that are half party and half showbiz, deeply patriotic military service, sacred places and secret ceremonies, the fraught opportunity of the casinos, many a prison, the endless jockeying with the Bureau of Indian Affairs, a prospering art and craft market, the revival of tribal languages, and count-less matters that a non-Indian like me has no inkling of. Native America is robust.

You can get a sense of its rootedness in specific landscapes with a stroll through Peter Nabokov's *Where the Lightning Strikes: The Lives of American Indian Sacred Places* (2006). The book chronicles the political battles to protect sacrosanct land features like Blue Lake, near Taos Pueblo, and the Black Hills, precious to Lakota, Cheyenne, and Kiowa. With his own personal visits, Nabokov demonstrates why such places are worth defending, and how many there are on the land:

A tally in 1980 by the California Indian Heritage Commission counted fifty-seven thousand old village, rock art, burial, food-

and-plant collecting, spirit homes and prayer sites located across 70 percent of the state.

"There is something to be learned from the native American people about where we are," wrote the poet Gary Snyder. "It can't be learned from anybody else."

Nearly every nation in the Americas has intact native populations. Some European countries do, such as Sweden with the Sámi (formerly known as Lapps). Australia has its Aborigine people, and New Zealand the Maori. Part of Africa's wealth is the most sustained human continuity in the world. Most Asian nations have some people who have been there the longest, such as the Ainu in Japan. The native traditions of the most experienced and attuned dwellers of the land are worth preserving and reviving everywhere.

• American Indians and the non-Indian majority—and the landscape we now share—are still getting over the trauma of the tectonic collision in the sixteenth and seventeenth centuries between shipborne Europeans and the indigenous peoples of the Americas. A continental population estimated to have been between 50 million and 100 million in 1491 was reduced to 6.5 million by 1650. It was the greatest cataclysm in human history; a fifth of the world's population died. We think of it as a military event, but it was almost entirely biological.

With no immunity to European diseases, native populations were shattered by successive waves of smallpox, influenza, hepatitis, whooping cough, dysentery, diphtheria, measles, mumps, cholera, typhus, yellow fever, scarlet fever, and bubonic plague. Densely populated regions like the Mississippi Valley, the Amazon, and the Yucatán Peninsula were so emptied that European explorers who arrived after the diseases swept through thought that no one lived there. The Pilgrims only survived their first New England winter, in 1620, thanks to food caches they found in abandoned native villages. The following autumn, the local chief Massasoit could afford to be generous at the first Thanksgiving: Disease had reduced the confederation he led from twenty thousand to one thousand. There was food to spare.

The book to read on all this is Charles Mann's *1491: New Revelations of the Americas Before Columbus* (2005). Crammed with news of every sort, it is especially germane here as a primer on how to harness an ecosystem. Before the great dying, the American continent was a managed landscape. Afterward, it was an abandoned garden that the Europeans misinterpreted as wilderness. For a time there were parklike forests free of undergrowth, but the people who had kept the undergrowth down with fire were dead, and the woods became impenetrable. Millions of passenger pigeons darkened the skies because their human competitors for food were gone. Millions of bison overran the plains because so many of the Indians who had hunted them had died. "Far from destroying pristine wilderness," Mann writes, "Europeans bloodily created it."

No wonder the reforesting of the continent caused the drawdown of atmospheric CO_2 noted by climatologist William Ruddiman. Before Columbus arrived, according to Charles Mann,

> Agriculture occurred in as much as two-thirds of what is now the continental United States, with large swathes of the Southwest terraced and irrigated. Among the maize fields in the Midwest and Southeast, mounds by the thousand stippled the land. The forests of the eastern seaboard had been peeled back from the coasts, which were now lined with farms. Salmon nets stretched across almost every ocean-bound stream in the Northwest. And almost everywhere there was Indian fire.

● Geneticist Nina Federoff has described maize as "arguably man's first, and perhaps his greatest, feat of genetic engineering." She told Charles Mann: "To get corn out of teosinte is so—you couldn't get a grant to do that now, because it would sound so crazy. Somebody who did that today would get a Nobel Prize! If their lab didn't get shut down by Greenpeace, I mean." Besides converting an unpromising grass into the world's most efficient and popular food plant, the pre-Mexican Indians developed the milpa field system to grow it in, combining, Mann writes, "maize, avocados, multiple varieties of squash and bean, melon, tomatoes, chilis, sweet potato, jicama (a tuber), amaranth (a grainlike plant), and *mucuna* (a tropical legume)." The beans climbed up the corn stalks for sunlight, fixed nitrogen in the soil, and provided dietary niacin that corn lacks. Add the vitamins in squash and the fat in avocados, and you need very little meat for a healthy diet.

Corn and the milpa made the high civilizations of Mesoamerica possible. "In the Mayan creation story, the famous *Popul Vuh*," writes Mann, "humans were literally created from maize." (Michael Pollan makes a similar point, less approvingly, in his exposé on American agriculture, *The Omnivore's Dilemma*.) Sustainability was built in—some milpa fields in Mexico have been in continuous cultivation for four thousand years. The many contemporary proponents of polyculture farming are heirs to milpa-style sophistication developed by ancient societies all over the world.

Recent studies of the Amazon have turned up another spectacular early agricultural invention. The basin's rain-forest soil is so infertile that everyone assumed only a scattering of hunter-gatherer tribes could live there, because agriculture would be impossible. That theory held sway despite accounts from sixteenth-century Spanish explorers of the Amazon reporting "numerous and very large settlements" and "one town that stretched for 15 miles without any space from house to house." In the 1990s, archaeologists began discovering extensive earthworks in the rain forest that indicated previous dense inhabitation, along with areas of a peculiar dark soil called in Portuguese *terra preta do Índio*—black earth of the Indians. Intensely and durably fertile, *terra preta* was generated apparently on purpose over large areas by an agricultural technique now called slash and char. The barren rain-forest soil was being amended with man-made charcoal. "Faced with an ecological problem," Mann writes, "the Indians fixed it. Rather than adapt to Nature, they created it. They were in the midst of terraforming the Amazon when Columbus showed up and ruined everything."

These days a *terra preta* craze has gripped bioenergy experimenters and agricultural scientists, who call the miracle substance *biochar*. Once in the soil, the charcoal provides habitat for microorganisms, retains soil nutrients, and sequesters carbon, apparently for thousands of years—the oldest sites date back 4,500 years. It reduces nitrous oxide emissions and prevents runoff of phosphorus and nitrogen in the rains. Some *terra preta* soil goes six feet deep, and it typically retains ten times as much carbon as ordinary Amazon soil. At biochar conferences, businesses, and do-it-yourself Web sites, there is talk of a "black gold revolution" in agriculture and a "carbon-negative bioenergy industry." An article in *Discover* enthuses:

> Burning agricultural wastes in a controlled process called pyrolysis can convert wood and other organic waste into useful volatile gases, heat, electricity, and bio-oil. The process is win-win: Burning the biomass produces substantial amounts of rich biochar from waste material like peanut shells and rice husks, and mixing this biochar into soil could more than offset the carbon that is emitted into the atmosphere not only during the burning process itself, but also when the derived fuels are used.

At one biochar conference, Australian biologist Tim Flannery declared that "biochar may represent the single most important initiative for humanity's environmental future. The biochar approach provides a uniquely powerful solution, for it allows us to address food security, the fuel crisis, and the climate problem, and all in an immensely practical manner." Needless to say, the fact that Indians discovered the benefits of biochar first makes the whole subject extra cool.

• If *1491* is the best introduction to indigenous ecosystem engineering, the most illuminating textbook is *Tending the Wild: Native American Knowledge and the Management of California's Natural Resources* (2005), by ethnoecologist Kat Anderson. The Spanish and then the Yankees who came to California in the eighteenth and nineteenth centuries "did not find in many places a pristine, virtually uninhabited wilderness," she writes, "but rather a carefully tended 'garden' that was the result of thousands of years of selective harvesting, tilling, burning, pruning, sowing, weeding, and transplanting." There were grasslands and meadows, enormous trees and parklike forests, spectacular wildflowers, and an abundance of game that were the unrecognized product of careful tending by the region's five hundred tribes. Two centuries later, a Southern Sierra Miwok elder, James Rust, told Anderson, "The white man sure ruined this country. It's turned back to wilderness."

California Indians were particularly expert in the use of fire, Anderson writes:

> Deliberate burning increased the abundance and density of edible tubers, greens, fruits, seeds, and mushrooms; enhanced feed for wildlife; controlled the insects and diseases that could damage wild foods and basketry material; increased the quantity and quality of material used for basketry and cordage; and encouraged the sprouts used for making household items, granaries, fish weirs, clothing, games, hunting and fishing traps, and weapons. It also removed dead material and promoted growth through the recycling of nutrients, decreased plant competition, and maintained specific plant community types such as coastal prairies and montane meadows.

California Indian women crafted the finest and most beautiful baskets in the world, using seventy-eight species of plants. Cooking the region's staple food, acorn mush, involved dropping hot rocks into baskets so tightly woven they held water. Materials that fine required serious horticulture. According to Anderson, "almost every type of sedge, wildflower, fern, bush, tree, or grass used in a basket was fussed over and meticulously groomed by weavers." Women would have done that work, because "in aboriginal California, women were the ethnobotanists, testing, selecting, and tending much of the plant world, and men were the ethnozoologists, applying their intimate knowledge of animal behavior to skillful hunting, fishing, and fowling."

Anderson learned that there were strict rules: Do not waste what you have harvested, and do not harvest everything. You leave some for the other animals; you leave some to grow back. This year, take your roots from one side of the tree only, then from the other side next year. (The Hopi Thomas Banyacya, who advised me never to collect herbs from the first bush, was making a similar point.) There were two levels of ownership rights—private to an individual or family, and communal to all. With the Pomo, one anthropologist found, "all the great oaks in the valley flat were privately owned; those of the hills were owned by the village as a whole."

"Take care of nature, and it will take care of you," a Pit River gent named Willard Rhoades told Kat Anderson.

✦

The delectable term *ecosystem engineer* was introduced in 1994 by ecologist Clive Jones and then was connected to the new ideas of *niche construction* and *ecological inheritance* by Oxford bioanthropologist John Odling-Smee (and coauthors) in 2003. Beavers perform niche construction when they create ponds with their dams, as do earthworms when they remake soil to suit themselves. In the process, both make richer environments for other organisms, and that qualifies them as ecosystem engineers. Because the transformed environments persist, that "ecological inheritance" becomes an evolutionary path that supplements genetic inheritance. The beavers and worms adapt progressively to the niches that they construct.

" 'Evolution helps those who help themselves' is the basic idea behind the concept of niche construction or ecosystem engineering," wrote archeobiologist Bruce Smith in a 2007 *Science* paper he titled "The Ultimate Ecosystem Engineers"—by which he means us. It's part of our very origin. Evidence for controlled burning by early humans, he notes, dates back fifty-five thousand years in southern Africa.

Smith offers an interesting angle on domestication: "In Asia, . . . the domestication of two utilitarian species—the dog (for hunting) and the bottle gourd (for containers)—by 12,000 years before the present, did not so much involve deliberate human intervention as it did allow dogs and bottle gourds to colonize the human niche." From this perspective, we didn't shape dogs; we allowed them to shape themselves to ride along on our landscape-tuning success story.

● One of the finest examples of beautifully nuanced ecosystem engineering is the thousand-year-old terraced rice irrigation complex in Bali.

"Perfect order" is how anthropologist Stephen Lansing describes it in his talks and books analyzing the Balinese water temple system. The elaborate array of ninth-century tunnels and rice paddy terraces in steep volcanic mountains is managed from the bottom up through a system of what are called *subaks*. Each *subak* is a group of men with adjoining rice fields that share a water source. Their meetings are democratic, dispensing with the otherwise powerful caste distinctions of Bali, and *subaks* are connected to one another through the hierarchic water-temple network. "The Balinese call their religion Agama Tirtha—the religion of water," says Lansing. "Each village temple controls the water that flows into nearby rice terraces. Regional water temples control the flow into larger areas."

The universal problem in irrigation systems is that upstream users have all the power and no incentive to be generous to downstream users. They can just keep all the water and put downstream competitors out of business. What accounts for their apparent generosity in Bali? Lansing discovered that the downstream users also have power, because pests can be controlled only if everybody in the system plants rice at the same time (which overloads the pests with opportunity in one brief season and starves them the rest of the time).

If the upstreamers didn't let enough water through, the downstreamers could refuse to synchronize their planting, and the pests would devour the upstreamers' rice crops. That balance of power is enough to enforce perpetually renewed agreement among the *subaks*. The system is kept legitimate by the democratic transparency of the governance, and, Lansing says, "the water temple rituals are intended to tame the passions and keep order." In Balinese language and understanding, he says, "rice paddies equals 'jewel' equals 'mind.'"

The green revolution (indeed playing the villain this time) arrived in Bali in 1971, funded by the Asian Development Bank. For the new program, the government provided "technology packets" of fertilizer, pesticides, and pest-resistant seed and urged the farmers to "plant as often as possible" instead of on the old pulsed schedule of the water-temple system. The result: Millions of tons of rice were lost, year after year, mostly to voracious pests. The level of pesticide use kept being increased, to ever-decreasing effect. Thanks to Lansing's research, the bureaucrats at the Asian Development Bank became converts to the water-temple system, and in the 1980s, the Balinese government was persuaded to throw out the green revolution pesticides and the "plant often" directive. Balance returned.

With the system functioning again, the *subaks* have gotten good value from the new higher-yield rice varieties. Unfortunately, the farmers continue to load their fields with fertilizer—needlessly so because Balinese water is naturally rich in nutrients. All that superfluous fertilizer passes through the watershed out to the sea, where it is destroying the coral reefs with suffocating algal blooms—a process known as eutrophication. The fishermen who are harmed by that are Bougainese rather than Balinese, and so their complaints don't connect back upstream. The rice farming may stay healthy for another thousand years, but the coral reef fishes are in trouble.

Bali's water-temple system presents an exquisite case of managing the commons, of refining and maintaining natural infrastructure. But what about the coral reefs? One hope may be the dive industry for ecotourists: It is run by Balinese, and they could put pressure on the government to discourage farmers from overloading the waters with fertilizer. For really perfect order, the dive operators should join the water-temple religion and organize their own *subak*.

● Ecosystem engineering can be beneficial or pathological. It can enhance biodiversity or decrease it. It can self-stabilize or go chaotic. The classic pathological case is what humans have done to most of the world's large animals—the megafauna. As soon as we arrive in virgin land, we kill and eat or burn out everything that is big or slow: in Australia, the enormous lizards, wombats, giant kangaroos, and flightless birds; in Madagascar, the giant hippos, mongooses, lemurs, and tortoises; in North America, just thirteen thousand years ago, the mastodons, mammoths, sloths, tigers, and yaks. Three fourths of our continent's large animal species disappeared in what are called the Pleistocene extinctions. As conservation biologist Daniel Janzen put it at a talk in California,

"There were animals walking around here that weighed twice what an African elephant weighs. They're still down there in the La Brea tar pits. Our people took them out. It's like going to East Africa and machine-gunning the elephants."

Most consequentially for later history, the first people in the Americas extirpated the native horses, camels, oxen, and pigs—animals they might have domesticated. And so when Europeans arrived after 1492, they brought not only a cavalry advantage in warfare but also lethal diseases—evolved from long cohabitation with domesticated cattle, pigs, and chickens in Eurasia—to which they were relatively immune. The American continent that had originally been tamed with spears, fire, and plant-tending women was overrun by a new set of ecosystem engineers armed with guns, germs, and steel.

Human prehistory confronts us with two harsh truths. Edward O. Wilson, who coined the term *biodiversity*, states one: "Humanity has so far played the role of planetary killer, concerned only with its own short-term survival. We have cut much of the heart out of biodiversity."

Steven LeBlanc states the other truth in *Constant Battles*:

> If any group can get itself into ecological balance and stabilize its population even in the face of environmental change, it will be tremendously disadvantaged against societies that do not behave that way. The long-term successful society, in a world with many societies, will be the one that grows when it can and fights when it runs out of resources. It is useless to live an eco- logically sustainable existence in the Garden of Eden unless the neighbors do so as well.

That's the pattern. Humans invade new places, wreak destruction, then eventually settle down. They invade each other, slaughter, then eventually settle down. Lately we've run out of unpopulated places to invade, and we say we'd like to stop invading each other. If we can keep climate change from forcing us back into the invasion game, how would settling down once and for all work? The encouraging lesson from prehistory is that settling down means inhabiting a place with such close attention that the practices of "tending the wild" and agriculture blur into one.

✦

"It is imperative that we come to recognize conserved wildlands for the gardens that they are," ecologist Daniel Janzen wrote in an essay for America's *National Parks* magazine. "They are water factories, amusement centers, grocery stores. They are the globe's finest research, entertainment, and aesthetic living libraries. They are carbon deposits, biodegraders, recyclers, buffers, ameliorators. They are sandboxes and swing sets. And they have shoplifters, vandals, drunks, speeders, and stupidities."

Janzen speaks with authority. He and his wife, biologist Winnie Hallwachs, are credited with helping to set Costa Rica on its path as the world's Greenest nation: The whole country considers itself a national park. To build the famed Guanacaste Conservation Area, now a World Heritage Site, "Janzen and Hallwachs assembled a set of ecological, political, cultural, and financial insights that, used together, constituted a revolution in tropical forest conservation," writes William Allen in *Green Phoenix*, a 2001 history of the project.

Janzen's experience in the tropics around the world told him that "protected" wildlands can't be protected if the only justification is aesthetic or political. Poachers, loggers, and new political regimes turn the parks into battlefields, and that means annihilation. To protect a wilderness permanently, he insists, it has to be treated not just as a garden, but as a *commercial* garden. The wildland must pay its way or perish.

A wilderness park like the Guanacaste, Janzen wrote in *Science*, has "nearly all the traits that we have long bestowed on a garden—care, planning, investment, zoning, insurance, fine-tuning, research, and premeditated harvest." He considers the 235,000 identified species in the Guanacaste as crops; ecotourists, as "a better kind of cattle." Janzen points out that "the ecotourist crop in Costa Rica is worth more than the combined coffee crop, banana crop, and cattle crop together, and Costa Rica is the number-two banana producer in the world." Janzen specializes in finding ways to get financial yield for the Guanacaste from scientific research, from drug bioprospectors, and from contracts for sundry ecosystem services, such as the fresh water that flows out of the protected forest. When the growing park displaces ranching and logging, he makes sure the local people have new and better economic opportunities in and around the park, often as parataxonomists and para-ecologists doing world-class science.

(It annoys Janzen and Costa Ricans when writers like me suggest that one outsider turned a whole nation around. Janzen was the right grain of sand in a very intelligent oyster. According to *New Scientist*, Costa Rica is "the first tropical nation to reverse deforestation. Thanks to conservation and replanting, its forest cover has increased from 21 per cent in 1987 to 52 per cent today." Costa Rica did that, not Professor Janzen at the University of Pennsylvania.)

● The benefits of commercializing the wild can be found wherever you look closely. For example, *Nature* reported that "whale-watching is already employing more people and bringing much greater economic benefits than commercial whaling could ever generate." The Marin County mountain I hike every week is a patchwork of national park, state park, county open land, and the Marin Municipal Water District. Guess which of those is the best maintained, with the best trails, the best volunteer programs for habitat restoration, the fewest alien-invasive species, and the most reliable decade-to-decade funding. Of course it is the 21,000-acre water district. Its 190,000 customers, including me, gratefully pay the district $55 million a year to protect the watershed. The toll collected on the water we drink pays for the trails we hike. The trees that shade and hold the soil and the mountain lions who manage the deer population are indirect beneficiaries.

Peter Kareiva, chief scientist for The Nature Conservancy, articulates a growing realization among environmental organizations: "We have to stop thinking of protected areas as 'protected from people' and recast protected areas as resources and assets that are 'protected *for* people.'" In a 2007 *Scientific American* article cowritten with Michelle Marvier, Kareiva gives an example of that approach. Florida's Gulf Coast is vulnerable to storm surges and is losing species-rich marsh and oyster-reef habitat.

Nature Conservancy researchers layered maps of the most critical habitats, the densest human populations, and the greatest storm hazard on top of each other and looked for overlaps. Then they directed their effort to the places where habitat protection and restoration could protect the most humans as well as the most wild species. Kareiva urges that "conservationists focus foremost on regions where the degradation of ecosystem services most severely threatens the well-being of people: stands of mangroves in Asia, marshes in the southeastern US, drylands in sub-Saharan Africa, and coral reefs around the world."

Daniel Janzen argues that you cannot protect a wildland by just leaving it alone. He wrote in *Science:*

> The question is not whether we must manage nature, but rather how shall we manage it—by accident, haphazardly, or with the calculated goal of its survival forever? . . . Restoration is key to sustainable gardening. Restoration is fencing, planting, fertilizing, tilling, and weeding the wildland garden: succession, bioremediation, reforestation, afforestation, fire control, prescribed burning, crowd control, biological control, reintroduction, mitigation, and much more.

A major attraction of the restoration approach is that it brings many people into the process. In *Nature by Design: People, Natural Processes, and Ecological Restoration* (2003), Eric Higgs points out that "the act of pulling weeds, planting, configuring a stream bank to match historical characteristics, or participating in a prescribed fire that returns an old process to the land helps develop a ferocious dedication to place." The volunteers bond with the land and with each other, and they make a deep connection to time and history. As Higgs writes, "We restore by gesturing to the past, but our interest is really in setting the drift pattern for the future." I would add that restoration volunteers also learn humility, because they are joining what Michael Pollan calls "the mostly comic dialogue with other species that unfolds in the garden." You diligently weed and plant, and then watch the vegetation do something quite different from what you had in mind.

In this century, global warming is forcing preservationists to completely rethink what they do, because their goal of preventing change to an ecosystem is no longer possible. Their job now is to help the ecosystem adapt. To stay in a viable temperature range as the climate warms, species are moving to higher altitudes and higher latitudes. They need protected corridors to do that. Where there's a barrier, such as a mountain range or body of water, they may need help getting across it. Species once thought of as alien invasive may need to be welcomed; new spontaneous hybrids such as "grizzlars" (a cross between a grizzly and a polar bear) that used to be prevented now should be supported as adaptive.

Populations in some locations, such as near mountaintops, may be impossible to protect; they will have to be abandoned. At the same time, preservationists are scouting for future opportunities. Peter Kareiva at The Nature Conservancy notes that "rising sea levels may turn today's nondescript inland habitats into tomorrow's valuable marshes and wetlands."

● This chapter makes so emphatic a case for tending the wild, for people being densely involved with nature, that I need to include the opposite approach—the leave-it-bloody-well-alone option. In all of Europe, there is just one remnant of truly ancient primary forest. The Puszcza Białowieska, all 380 square miles of it straddling the border between Poland and Belarus, is a World Heritage Site, a Biosphere Preserve, and a national park in both countries. "Here," writes Alan Weisman in *The World Without Us* (2007), "ash and linden trees tower nearly 150 feet, their huge canopies shading a moist, tangled understory of hornbeams, ferns, swamp alders and crockery-sized fungi. Oaks, shrouded with half a millennium of moss, grow so immense here that great spotted woodpeckers store spruce cones in their three-inch-deep bark furrows."

Weisman's guide, a Polish ecologist named Andrzej Bobiec, drew a lesson from the forest that is radical in Europe:

> As a forestry student in Krakow, he'd been trained to manage forests for maximum productivity, which included removing "excess" organic litter lest it harbor pests like bark beetles. Then, on a visit here he was stunned to discover 10 times

more biodiversity than in any forest he'd ever seen. It was the only place left with all nine European woodpecker species, because, he realized, some of them only nest in hollow, dying trees. "They can't survive in managed forests," he argued to his forestry professors. "The Białowieza Puszcza has managed itself perfectly well for millennia."

I heard a similar tale from Jim Lovelock about the Devon farmstead he bought thirty years ago: "When I started this place, I wanted to be good and Green, so I planted twenty thousand trees. It was a mistake, really. They haven't gone all that well. But the few bits of land that I left untouched, just left to their own devices, are now flourishing ecosystems, with trees and everything else in them. My whole philosophy now is: Leave it alone, don't touch it. It's easier, too."

Where do you suppose is the richest wildlife refuge in northeast Asia? Here's a hint: it contains (according to an online source):

five rivers and many ecosystem types, including forests, mountains, wetlands, prairies, bogs and estuaries. There are over 1,100 plant species; 50 mammal species, including Asiatic Black Bear, leopard, lynx, sheep and possibly tiger; hundreds of bird species, many of which, according to IUCN, are endangered, including Black-faced Spoonbill, Red-crowned and White-napped Cranes and Black Vulture; and over 80 fish species, 18 being endemic. . . . Hundreds of bird species migrate through.

(IUCN stands for the International Union for Conservation of Nature, which maintains the authoritative Red List of Threatened Species.)

Of course it is the DMZ—demilitarized zone—separating North and South Korea: 155 miles long, 2.4 miles wide, full of mines and empty of people ever since the armistice of September 6, 1953. Forests grew back in the mountains. Ancient agricultural land returned to prairie and shrubscapes. No one had to introduce wildlife: As at Chernobyl, the animals found their own way to the inadvertent refuge. (An NGO called the DMZ

Forum is lobbying to convert the DMZ to a peace park when the two Koreas eventually reunite. Kept as wildland, it could be a perpetual source of national pride and ecotourist revenue. "Think of it as a Korean Gettysburg and Yosemite rolled together," says Edward O. Wilson, cofounder of the DMZ Forum.)

Similar wildlife sanctuaries can be found wherever there's barbed wire and paranoia, such as around nuclear plants. "Wildlife is not something most people associate with national labs and military bases," says preservation lobbyist Corry Westbrook, "but restricted access and high security make them ideal spots for conservation." These areas are a survival into the modern age of the timeless practice of no-go buffer zones between competing tribes, as described in Steven LeBlanc's *Constant Battles*. Typically ranging from a half mile to twenty miles in width, the buffer zones maintained to reduce combat were fortuitously a reservoir and corridor for game animals and a place where worn-out agricultural land could lie fallow while its soil revived.

Tending the wild and leaving the wild alone are not contradictory strategies. They both work fine, and they blend well. In any project, it's a good idea to try some of both, in different areas, so that each is a scientific control for the other. Race them.

✦

Restoring the wild is the most rewarding of all Green activities, at every scale from window box to biome. Let me give some examples from cities, agricultural lands, forests, and prairies before moving on to the really large scale.

Thanks to a fluke of bridge engineering, the Congress Avenue Bridge in Austin, Texas, is home to 1.5 million Mexican free-tailed bats. Every summer evening, as a thousand tourists and locals look on, the bats swarm into the twilight to devour their nightly twelve tons of mosquitoes and agricultural pests such as corn earworm moths. An Austin totem animal of sorts, the free-tails earn the city an estimated $8 million a year. Bat Conservation International, based in Austin, has produced a publication showing how to make any bridge bat friendly.

Recent ecological studies in American suburbs have turned up three reasons for the alarming decline in wild bird populations. As I mentioned before, one is cats, who kill 100 million birds a year in the United States. Conservation biologist Michael Soulé discovered that in places where coyotes are on the increase, birds are returning as well, because the coyotes eat or intimidate the cats. Meanwhile, researchers in Pennsylvania found that wherever there are more than twenty deer per square mile, the bird population goes down by a third, because the excess deer eat up the understory where the birds live (along with chipmunks, squirrels, and other small mammals). Again the problem is the absence of large predators. We need bird-watchers and gardeners with a taste for venison. A number of cities now offer licenses for urban deer hunting with bow and arrow. A third cause of bird decline is the popularity of "pest-free" exotic plants in gardens and yards. Local insects haven't learned that the alien plants are food; so the insects dwindle, and the birds starve. The solution there is simple: Plant natives.

That's pretty rewarding. If you tolerate the singing of coyotes, eat the beast that vaults an eight-foot fence to devour your garden, and rid your yard of plasticlike ornamentals, you get the birdies back.

● Environmentalists are used to thinking of cattle as the great destroyers of land, and indeed they often are. But if they're managed right, livestock can be grassland preservers, standing in for the missing megafauna. The trick is to keep the animals moving, with cowboys replacing the large predators of old. The guru of enlightened cattle ranching, biologist Allan Savory, realized from field work in his native Zimbabwe that "relatively high numbers of heavy, herding animals, concentrated and moving as they once did naturally in the presence of predators, support the health of the very lands we thought they destroyed."

Savory's now-classic book, *Holistic Management* (1999), shows how to employ grazing animals as tools of land restoration. Got some bare ground that's eroding away? Concentrate the herd on it with bales of hay; once the cattle have broken up and dunged the ground, the plants will come. Got a fast-eroding gully with steep banks? Let the herd flatten out the banks;

vegetation will finish the work. Need a firebreak through a brushy area? Spray some dilute molasses or salt solution on the strip you want cleared, and the cattle will bulldoze your firebreak.

Among the ranchers paying attention to Savory are the one hundred member families of the Malpai Borderlands Group, which oversees 1,250 square miles of wild country on the Continental Divide in southern New Mexico and Arizona. The ranchers came together in the 1990s to figure out a way to restore fire to their landscape, because woody plants were taking over the grasslands. That led to bringing in scientists and figuring out other projects they could collaborate on. Pretty soon a book about the Malpai Group described what they have as

> a *working wilderness:* a place where wildness thrives not in the absence of human work or in spite of it, but because of it, and where thriving wildlands in turn sustain the human community that lives and works there. Superficially, this work is the same as it has been for five generations—the work of ranching, of raising cattle on the range—but it has grown to include scientific research, communications and outreach, real estate, law, wildlife biology, planning, and fire management as well. Not to mention politics.

The Malpai Group hired as a consultant the African-born conservationist David Western, famous for protecting Amboseli National Park by allying its interests with those of the surrounding Masai cattle herders. Then head of the Kenya Wildlife Service, Western arranged for the American ranchers and the Masai to visit each other. For a further thrill, now that Mexican jaguars have been sighted in the Malpai area, the group is working to restore that spectacular animal to at least part of its former range in the American southwest.

Imagine how satisfying it is for ranchers who used to be harassed by government officials and scorned by environmentalists to now have both coming to study their techniques. Plus, they get to party with the neighbors, play with fire, identify with jaguars, and hang out with scientists and

Masai warriors, all while improving their grasslands and watercourses and selling ever more grass-fed beef.

● Forests have often been full of people. The medieval forest of Europe, Roland Bechman writes in *Trees and Man* (1990), was populated with "basket-weavers, charcoal-burners, hoop-makers, potters, loggers, etc., as well as shepherds, herdsmen, and swineherds." By studying satellite images of the dry forest in Madagascar—one of the world's most important ecological areas—Thomas Elmqvist of Stockholm University discovered that while much of the forest was being lost, in some places it was intact and even increasing. "We were surprised to find that areas that were suffering most from deforestation had the lowest population density and were far from markets," he told *New Scientist*. The robust areas of forest were protected by their local villages: "If an outsider wants to use the forest, the only way to get permission is to marry into the clan."

Because of climate concerns, forests are now seen as crucial for their role in fixing and retaining carbon. The Intergovernmental Panel on Climate Change says that converting 2 billion acres of farmland to agroforestry (which integrates trees, shrubs, livestock, and row crops) would remove 50 gigatons of carbon dioxide from the atmosphere. The World Agroforestry Centre suggests that "allowing farmers to sell that carbon on global carbon markets could generate as much as $10 billion each year for poor people in rural areas."

Forests change climate, and climate changes forests. In *America's Ancient Forests* (2000), Thomas Bonnicksen offers a bracing perspective:

> Modern forests only exist today. They do not look like Ice Age forests nor do they look like forests of the future. Forests represent a loose collection of species that grow together for a time as they pass each other on their way somewhere else. Each species arrives and departs independently from other species. Plants move very slowly; animals move more quickly; but they all continue to move either to escape an inhospitable

environment or to take advantage of a new one. If they cannot move, they adapt. If they cannot adapt, they become extinct.

(That's what happened to the sweet old idea of "ecological community"—everything is just "a loose collection of species," densely related for a while, but devoid of cohesion.)

From his experience restoring cloud forest and tropical dry forest in Costa Rica, Daniel Janzen developed a formula that can be applied to any forest revival:

> Choose an appropriate site, obtain it, and hire some of the former users as live-in managers. Sort through the habitat remnants to see which can recover. Stop the biotic and physical challenges to those remnants. The challenge is to turn the farmer's skills at biomanipulation to work for the conservation of biodiversity. Explicit and public agreement on management goals is imperative. Is the goal a low-overhead zoo, botanical garden, gene bank, functioning watershed, teaching laboratory, or some combination of these and other goals?

In Borneo, which has been severely deforested by logging and fires, two forest restoration projects with interestingly different goals are under way. One is structured around the idea of rebuilding habitat for endangered orangutans. Working closely with six hundred Dayak families, forester Willie Smits is transforming palm-oil plantation land back into rain forest, employing a successional strategy accelerated with intense use of compost. The initial eight-square-mile forest is surrounded by a ring of fire-resistant, income-producing sugar-palm farms. The second grand scheme, called the Planted Forest Project, seeks to establish truly sustainable logging in Borneo. Half of its 1,900 square miles will be devoted to fast-growing acacia tree plantations, and a third to wild forest linked by corridors for wildlife, leaving the rest for the resident indigenous people. How the two projects fare over the coming decades will be fascinating to compare.

The Scottish highlands, now famously barren, once were covered with a dense forest of Scots pine, birch, juniper, rowan, and alder. In the mid-eighteenth century, when Scotland's last wolves were killed, the red deer population exploded and devoured all the tree seedlings. Only a few tiny remnants of the old Caledonian forest survived. Now a group called Trees for Life is working to reforest six hundred square miles. In 2008 they acquired a ten-thousand-acre estate, where volunteers are based for restoration chores on the property and in nearby forest remnants. The next stage is to reintroduce wild boars and beavers, absent for centuries. The beavers, if they're allowed to, will engineer the Scottish ecosystem in their customary fashion, and animal and plant diversity will increase greatly.

• Ecotrust—a Portland-based conservation group that distills its aims with the words "economy, ecology, equity"—is attempting the most ambitious forest-economy project I know of, embracing the entire temperate rain forest that extends two thousand miles up the Pacific coast from San Francisco to Alaska. Ecotrust's president, Spencer Beebe, describes it:

> There is a 5-million-acre marine-terrestrial system unlike any other, where just 7 percent of the landscape is in private industrial ownership. We can make 6–8 percent real rates of return by buying the logged-over land and committing to an investment model that layers ecosystem service revenues (water, carbon, cellulosic ethanol, mitigation banking, conservation easements, etc.) on top of product sales (pulp and sawlogs through thinning practices which mimic natural disturbance patterns).

Ecotrust calls the region Salmon Nation, for the charismatic species that has long served the area as a traditional native food source and major export and is considered the most sensitive indicator species of the ecosystem's health. (I'm an adviser to Ecotrust, partly because of my abiding love of Oregon. I did my major tree-hugging there in 1957 as a logger. Choker-setters hug trees for a living, under circumstances where the log can hug you back—squish you flat. I learned an important fact that escapes most

Greens: Loggers love the woods. That's why a vision like Beebe's might work.)

Magnificent chestnut trees once dominated the forest that cloaked the continent from Alabama to Maine, from the Mississippi to the Atlantic. They grew to 130 feet high with trunks ten feet in diameter, and they rained down an annual harvest of sweet nuts that fed humans, squirrels, deer, elk, turkeys, bears, jays, and mice through the winter. In 1904 an invasive fungus began to kill them. By the 1920s, they were all gone, impoverishing the landscape. Heroic efforts to produce a resistant variety by crossbreeding with the Chinese chestnut (which is immune to the blight) failed utterly.

In Susan Freinkel's lovely book on the subject, *American Chestnut: The Life, Death, and Rebirth of a Perfect Tree* (2007), the rebirth refers to efforts to genetically engineer a blight-resistant American chestnut. Two researchers made a derivative of a frog gene they thought would do the trick, but everyone told them they mustn't put a frog gene in a plant that people eat. A forest biotech company named ArborGen approached the American Chestnut Foundation, offering to support research on a GE chestnut. They were turned away, of course. But research is going ahead anyway; I bet that an American chestnut 2.0 will be thriving by the 2020s and that Greens will welcome it by the end of that decade.

In *Intervention: Confronting the Real Risks of Genetic Engineering and Life on a Biotech Planet* (2006), Denise Caruso frets that "one concern is that the transgenic tree would itself become an invasive species." Actually, that's the idea. *Invasive* is a neutral term in ecology, meaning only that something is on the increase. If chestnuts take off with human help, it won't be the first time. About two thousand years ago, chestnuts suddenly jumped from 7 percent of the deciduous forest to 40 percent—impelled, it is presumed, by native Americans gardening the woods.

● Perhaps the most radical restoration scheme of all is a plan to undo the Pleistocene extinctions and restore megafauna to their original keystone role in the North American ecosystem. The motivation is not just nostalgia for big animals. To deduce what happened before and after the extinctions, science journalist Yvonne Baskin studied the work of South African ecologist Norman Owen-Smith, who has been researching what he calls the "keystone herbivore" hypothesis.

Before the extinctions, Baskin writes in *The Work of Nature* (1997), "The pollen record indicates that a wide sweep of the continent from the Appalachians to the Rockies was covered by open, parklike woodlands where conifer and hardwood trees were interspersed with stretches of grass and wildflowers." Apparently that idyllic landscape was created by mammoths and saber-toothed tigers and the rest. Soon after they were killed,

> open glades would have filled in with unbrowsed shrubs and seedlings, crowding out the grasses and herbs. The ungrazed grasses in the clearings would have grown tall, fueling more intense and frequent fires that killed off the seedlings in their midst. The result, Owen-Smith suggests, was the conversion of those mixed savannas into distinct zones of dense forest and uniform prairies typical of the region today. Without megaherbivores to create and maintain a diverse patchwork of habitats, the populations of many medium-sized and smaller creatures would have been fragmented into isolated and shrinking pockets of suitable living space. With escape routes increasingly blocked by open prairie or dense forests, the small mammals would have been at the mercy of a shifting climate, chance disturbances, and even human hunters. Eventually most of these creatures went extinct, too.

In 1999 the originator of the Pleistocene overkill theory, Paul Martin, was inspired by a conversation with Kenya's David Western to propose bringing the big tuskers back onto the American landscape as an element of "resurrection ecology" for the continent. Martin elaborates on the idea in *Twilight of the Mammoths: Ice Age Extinctions and the Rewilding of America* (2005). He was moved by Western's description of the Amboseli Park elephants, who browse on trees and shrubs and continually swap places with Masai cattle, who graze on grass. Once the grass is grazed down, shrubs and trees take over; this attracts the elephants, who knock down and eat the woody plants, restoring the area to grassland suitable for cattle. Western described the result as "a patchy mosaic of grass, woods, and shrublands. That's the whole reason for the savannah's diversity."

Western observed that "with elephants and cattle transforming the habitat in ways inimical to their own survival but beneficial to each other, they create an unstable interplay, advancing and retreating around each other like phantom dancers in a languid ecological minuet playing continuously over decades and centuries." Import some elephants, Paul Martin thought, and "in the New World we can see if bison and elephants too will dance this minuet, to the benefit of the American range."

Pieces of herbivore restoration ecology are already in place. Laws now protect the wild horses and burros that wander the West. "Because horses evolved here, flourished for tens of millions of years, and vanished around 13,000 years ago," Paul Martin notes, "their arrival with the Spanish in the 1500s was a restoration, not an alien invasion."

Buffalos—American bison—were reduced by 1890 to just 500 animals. Now their population is back up to 500,000, and they're paying their way. There's a good market for their lean, tasty meat, and they offer some savings over regular cattle. "Since buffalo evolved on the prairies," writes Alice Outwater, "they are far hardier than cattle and can be raised without antibiotics, hormones, or artificial growth stimulants. . . . They survive temperatures that freeze cattle solid." Ted Turner has 50,000 buffalo on his ranches (and he's reinstating prairie dogs, bless him). Some fifty-seven Indian tribes in nineteen states are members of the Intertribal Bison Cooperative, which helps restore buffalo to tribal lands to promote "cultural enhancement, spiritual revitalization, ecological restoration and economic development." Montana State University has a Center for Bison Studies. As people move out of the high plains "buffalo commons," bison are moving back in.

● Paul Martin's rewilding vision joins a similar one that has been promoted since 1991 by Dave Foreman, a founder of Earth First! Foreman's version of rewilding, inspired by ecologist Michael Soulé, is based on keystone carnivores instead of herbivores. In *Rewilding North America* (2004),

Foreman writes, "Wolves, cougars, lynx, wolverines, grizzly and black bears, jaguars, sea otters, and other top carnivores need to be restored throughout North America in ecologically effective densities." Like many others, he is impressed by the results of reintroducing wolves to Yellowstone National Park in 1995. Elk had overrun the park, eating the riparian aspens, willows, and cottonwoods down to the point that beaver could no longer build dams, and wetlands were being lost. Now that wolves have reentered the picture, Foreman writes, "Elk have become elk again. They're awake! They're moving. They're looking over their shoulders. They aren't loafing in big herds in open river valleys." And the beaver are back.

Conservation biologist Josh Donlan has blended the Martin and Foreman visions, saying it is time to reverse the "Pleistocene overkill" with a "Pleistocene rewilding." As lead author (with Foreman, Soulé, Paul Martin, and others) of a 2005 paper for the *American Naturalist*, Donlon proposes to introduce "surrogate" replacements for the lost North American megafauna. African cheetahs, along with some African lions, would replace the long-vanished American cheetah that made our pronghorn antelopes so speedy. Bactrian camels, endangered in their own Mongolian range, would stand in for their ancient counterparts here. Our present wild horses and burros would be joined by Europe's primeval-looking Przewalski's horse and Asian ass (both endangered in their home territories). African and Indian elephants would take up where our mastodons left off. For a long period of research, the animals would be contained within very large fenced parks, which would pay part of their way with ecotourism.

One flaw in the plan is that while elephants can substitute for mastodons (they both browse on trees and shrubs), there is no animal that can really replace the grass-eating mammoths. Our only option is to revive the mammoths themselves, something that is rapidly becoming feasible. A 2008 article in the *New York Times* estimated that "a living mammoth could perhaps be regenerated for as little as $10 million." Researchers at Pennsylvania State University are sequencing the mammoth genome from samples of mammoth hair. New GE techniques developed by George Church at Harvard and Shinya Yamanaka at Kyoto University would allow skin from an elephant to be reprogrammed to an embryonic state, then injected with multiple mammoth genes until one had what was effectively a mammoth embryo that could be brought to term in a mother elephant. It appears that Neanderthals could be brought back to life by the same technique, but they belong in Europe. Other resurrectable candidates for rewilding North America are the saber-toothed tiger, the short-faced bear, the giant ground sloth, the giant beaver, and the armadillo-like glyptodon.

Too far-fetched? Ecologist Sergei Zimov has been assembling a Pleistocene park in northeast Siberia since 1989. In a sixty-square-mile fenced preserve he is introducing reindeer, moose, Yakut horses, musk oxen, and American bison, soon to be followed by saiga antelope, yaks, wolverines, Asiatic black bear, and Siberian tigers. Japanese and Russian scientists are collaborating on cloning the woolly mammoth and possibly the woolly rhinoceros for him. Zimov's theory, expressed in a 2005 essay in *Science*, is that the moss and forest tundra of the region is an artifact of the killing of the ecosystem's megafauna, and when they are restored, the grass-dominated "mammoth tundra-steppe" will return. His hope is that

> by learning how to preserve and extend Pleistocene-like grasslands in the northern latitudes, we could subsequently develop means for mitigating both the progress and effects of global warming. The amount of carbon now sequestered in soils of the former mammoth ecosystem, and that could end up as greenhouse gases if released into the atmosphere by rising global temperatures, surpasses the total carbon content of all of the planet's rain forests.

Back in North America, Dave Foreman also thinks at continent scale. His rewilding vision is based on the idea that "nature reserves must be big and connected." Existing parks, wildernesses, and roadless areas, he insists, need to be linked by protected corridors—four "continental mega-linkages" going up the Pacific mountain ranges, the Atlantic mountain ranges, the continental divide, and across the Arctic-Boreal far north. It strikes me that his three north-south corridors are worth establishing for climate reasons alone. The Malpai jaguars may need to get to Canada.

✦

Nothing is as instructive as a worthy enemy.

My preferred start time is well before dawn. I like to be on the mountain before first light, finding my way across the terrain with skills and delight left over from Army night patrols. Wearing oiled pants to handle the chaparral, I'll have a minimalist pack with breakfast makings and survival oddments, plus work gloves, a large serrated military knife, and a pick mattock. When I get to the work area, I'll brew some coffee, munch granola with yogurt, and watch the world take light. Who meets the dawn owns the day.

My specialty is pampas grass, an alien invasive from Argentina and Bolivia, flamboyantly ornamental with its high blond plumes, long ago escaped from nurseries and people's yards into the national park west of the Golden Gate Bridge. In fact, it's not really pampas grass but its evil twin, jubata grass—identical except that it reproduces asexually: Every one of the million or so seeds blowing off each plant is fertile, and they can travel for miles on the wind. My goal is to eradicate jubata from the watershed. My technique is violent personal combat, using knife and pick mattock to dismember and uproot a plant that can be as big as a Buick.

Each session provides a half day of productive toil: finding the remote places where the foe takes root, bushwhacking my way there, uprooting the invader, admiring the improved landscape, and moving on. This is what I do instead of thrashing uselessly in a gym or pounding around a track. It particularly promotes upper-body strength, which makes the older male proud and therefore happy. I find fighting smart alien plants more engrossing than going after a wily trout or noble stag. Invasive plants are intensely skilled at what they do. It took me years just to master jubata-fu.

I'm a biobigot, a native-plant nazi. We are legion. Every state in the United States has a native plant society. Search for native plant organizations on Paul Hawken's WiserEarth database and a thousand listings come up. Innumerable native-plant nurseries serve our mania. State by state we are persuading our highway departments to stop mowing the 12 million acres of roadside and median strips and to plant local native plants, "so when you're driving around Delaware, you know you're in Delaware, not in the tropics," as one proponent put it.

• I regard the native-plant movement as an entirely wholesome phenomenon, much like bird-watching in that it grounds people in their local terrain and turns them into para-ecologists, enriching science. And it does improve the health of ecosystems. Among the growing number of restoration and native-plant professionals, many have adopted as routine some practices and ideas that surprised me and might surprise most people. Let's pretend they are secrets and reveal them here.

Secret: Serious restorationists use herbicides. Sometimes the scale or perniciousness of an alien-invasive problem defeats every other method. Yellow star thistle, a demonic weed, is enveloping the American West the way kudzu blanketed the South; it covers 15 million acres in California alone. Inedible and impassible for wildlife and livestock, it also drives out other plants. The Nature Conservancy uses herbicides such as Milestone, Tordon, and Transline against yellow star thistle in places like Hell's Canyon, Idaho. Invasive cheatgrass and medusahead have made the West so flammable (a danger that climate change is exacerbating), that the Bureau of Land Management is planning to spray a million acres with a herbicide called Plateau. *High Country News* reported: "Many environmentalists fear collateral damage. The biologists and land managers on the front lines, however, seem nearly unanimous in saying that the threat posed by weeds outweighs that from herbicides."

In the national park where I uproot jubata grass, there are some remote stands so thick that I and other volunteers can't make a dent; Park Service work crews blast those plants with herbicide. Restoring native grassland is customarily done by nuking whole pastures with herbicide and then starting over with native seed. At the river property where I'm restoring upland habitat, I occasionally use glyphosate against a dozen species of invader.

Secret: Heroic clearing of a dense patch of invasive plants is worse than useless. The disturbed ground just invites the same weeds back, or some new ones. The right solution is the Bradley method of regeneration, named for Eileen and Joan Bradley, who invented it in Sydney in the 1960s. Specialists ridiculed their technique until it was tested and then adopted by Australia's National Trust in 1975; it has spread virally ever since. The sisters' strategy is to let the native plants do most of the work against the aliens. You start where there are fewest invaders, uprooting them gently, and you take out every kind of weedy exotic, lest they replace each other. Go away for a season and let that set of natives grow strong and aggressive. Next season, help the natives advance further on the main body of invaders with some more gradualist weeding. After a few years of sporadic minimal work, the problem is solved permanently. The sisters' posthumous book, *Bringing Back the Bush* (2002), is worth tracking down.

Secret: Alien invasives increase biodiversity. New Zealand is a famously invaded place: It has 2,065 native plants, but 2,069 alien plants have taken up residence. Brown University ecologist Dov Sax points out that New Zealand's biodiversity has doubled, at a cost of just 3 documented plant extinctions. "I hate the 'exotics are evil' bit," Sax told the *New York Times*, "because it's so unscientific." Extinctions are caused by alien *predators*, he argues, seldom by alien competitors. According to an article in *Permaculture Activist*, "Out of a total flora of approximately 6,000 vascular plant species, California has more than 1,000 naturalized exotics; yet fewer than 30 natives are known to have become extinct."

Secret: Some alien invasives are good or become good. "It's hard to imagine the American landscape," wrote Michael Pollan in *Second Nature* (1991), "without St.-John's-wort, daisies, dandelions, crabgrass, timothy, clover, pigweed, lamb's-quarters, buttercup, mullein, Queen Anne's lace, plantain, or yarrow, but not one of these species grew here before the Puritans landed." The picturesque golden hills of summer in California are a European artifact: Annual grasses brought by the Spanish largely replaced the native perennial bunchgrasses, whose deep roots kept them somewhat green throughout the year. The new grasses spread so rapidly that by the time the Spanish got to northern California, the oat grass they had introduced in the south had preceded them, and the Indians were already eating the oats. An article in *Bay Nature* contends that "the Mediterranean annual grasses are . . . now as much a part of California's grasslands as the native perennial grasses once were. The time is long overdue for an official naturalization ceremony."

Consider the much-loathed zebra mussel. A classic invader via ballast water, it came from the Black Sea to one of America's Great Lakes in 1985. By 1996 it had taken over all the Great Lakes and most of the rivers of the Midwest, including the Mississippi. Zebra mussels fasten to hard surfaces in such density that they sink buoys, clog water intakes, coat ships, and suffocate other shellfish; they are a monumental hassle. But Dov Sax (him again), writing with his former teacher, ecologist James Brown, wants us to know:

> There are two sides to the story. Eutrophication has plagued the Great Lakes for decades. But by filtering phytoplankton and other suspended material from the water column, the non-native zebra mussel has helped clean up Lake Erie and other parts of the Great Lakes, the Hudson River, and many more aquatic environments. These mussels are much more efficient at filtration than their native counterparts. Many birds feed on them, and the mussels' excrement provides habitat for a food chain anchoring a great diversity of species. Biologists credit the zebra mussel with restoring native grasses and fishes. Were it native, the zebra mussel would be hailed as a savior, not reviled as a scourge.

Secret: Alien invasives aren't a problem in the tropics. "Those fancy exotics don't make it over the garden wall," Daniel Janzen told me. "It really is a jungle out there." That's true only on continents, of course. Replete with empty niches, tropical islands are extremely vulnerable to invasion. Australia is the biggest of such islands.

Secret: Climate change favors weeds. Endemic species are often highly specialized to local conditions, with limited range. When conditions change, they are the first to go and are replaced by the feisty, talented travelers we know as weeds. Defending most endangered species at ground level is doomed to fail as the climate shifts. Also, global warming leads to more fires, and the successional growth following a fire is likely to be different than before.

Secret: Biocontrol usually works. Yes indeed, the mongooses that were introduced in Hawaii in 1833 to kill rats ate everything else instead and became an ecological cautionary tale, along with the infamous cane toads imported to Australia in 1935 to control cane beetles. The toad has turned into a nightmare invasive, overrunning the country, eating everything it's not supposed to, poisoning every creature that tries to eat it—crocodiles, snakes, dingoes, quolls (marsupial cats), and pet dogs. Bash one with a cricket bat and poison sprays into your mouth. But neither mongooses nor cane toads went through the kind of evaluation process now required for biocontrol agents, all of which are insects or smaller because they can be selected for close specificity to the target organism. Three kinds of weevil and two kinds of fly are being used with considerable success by The Nature Conservancy in Hells Canyon in their campaign against yellow star thistle. In the last hundred years, 350 biocontrol agents have been deployed effectively against 133 species of weed, with just 8 cases of a non-target species being harmed—and then never seriously. The California Invasive Plant Council's newsletter declares, "The ideal biological control agent works year after year, spreading throughout the range of the targeted invasive plant, finding the most hard-to-reach plants. . . . It keeps on working long after we have forgotten that we ever had a problem."

Prediction: Restorationists will welcome genetically engineered biocontrol. The world's mitten crabs and comb jellies and fire ants and snakeheads and Nile perch and brown tree snakes and gorse and knapweed and water hyacinth and kudzu are just too damaging and impossible to contain with present techniques. Biocontrol organisms genetically engineered for extreme specificity to the target species are the obvious solution. It's already happening. At Australia's Commonwealth Scientific and Industrial Research Organization, in a project that has been under way since 2002, virologist Jackie Pallister is engineering a ranavirus genome to undo the cane toad mistake. It works better than cricket bats.

◆

"Ecology needs to be a predictive science," Edward O. Wilson told me. At present, ecology is still limited to being an observational science because the observation isn't complete yet. Some 1.6 to 1.9 million species—no one knows the exact number—have been identified since Carl Linnaeus founded taxonomy in 1735. Estimates of how many species there are in the world range from 3 million to 100 million (not including the microbes). In other words, we're so ignorant, we don't know how ignorant we are.

Gardeners know all the relevant species in their garden; they keep things simple so they can. If we're going to garden the wild (and the world) responsibly, simplifying is not an option; we have to inventory all of life in order to really understand food webs, energy webs, biogeochemical cycles, seasonal and climatic changes, shifting population ratios: the full gamut of how life works. "Imagine doing chemistry knowing only one-third of the periodic table," says Terry Gosliner, a mollusc expert at the California Academy of Sciences.

I learned all this because in 2000 I got involved with a scheme we called the All Species Inventory. It was Kevin Kelly's idea. In a founding statement, he wrote: "If we discovered life on another planet, the first thing we would do is conduct a systematic inventory of that planet's life. This is something we have never done on our home planet. The aim of the All Species Inventory is simple: within the span of our own generation, record and genetically sample every living species of life on Earth." My wife, Ryan Phelan, had just sold a company, so we put some money into gathering the world's leading taxonomists and systematists in San Francisco for a meeting to decide whether a push to identify all life was useful and feasible. Ed Wilson hosted a follow-on meeting at Harvard a few weeks later. The universal message from the scientists was to go for it.

As the project took shape, I got to participate in species inventories in Costa Rica, in the Great Smoky Mountains National Park, and inside a wood rat. (Why inside a wood rat? As with the human microbiome project, we are learning the degree to which life lives on life. Carl Zimmer wrote in *Parasite Rex* [2000]: "There's a parrot in Mexico with thirty different species of mites on its feathers alone. And the parasites themselves have parasites, and some of those parasites have parasites of their own. . . . According to one estimate, parasites may outnumber free-living species four to one. In other words, the study of life is, for the most part, parasitology.") One of the scientists donated $1 million to All Species, but little further funding came, and our organization faded by 2004, although several allied operations have prospered.

• Ed Wilson wrote a much-read 2003 paper for *Trends in Ecology and Evolution* titled "The Encyclopedia of Life." The goal, he said, is to "put all the information that we get on species already known into a single great database, an electronic encyclopedia, with a page that's indefinitely extensible for each species in turn, and that would be available to anybody, any time, anywhere, single access, on command, free." In 2007 the MacArthur Foundation, the Sloan Foundation, and other sources funded the project, and it partnered with heavyweights—the Smithsonian, the Field Museum, Harvard, Woods Hole, the Missouri Botanical Garden, and the Biodiversity Heritage Library, which is busy digitizing 500 million pages of papers on species to blend in to the encyclopedia. The hope is that by 2017 almost all of the 1.7 million or so known species will be in the database. This will be, Daniel Janzen declared, "*the* window on the biodiversity of the world, for the world, by the world."

That "by the world" part is what's revolutionary. Taxonomy, like the rest of science and academia, is turning upside down. The power has shifted from sequestering data to sharing data. A taxonomist used to build a career by taking ownership of a particular twig of the tree of life—some genus of beetles, say—and then controlling all information about that twig. If you discovered what you thought might be a new species in that genus, you sent your specimen to the specialist and then waited months for a verdict. The revolution started by GenBank changed all that. Because it was the freely accessible online repository of all genetic data, including the human genome, scientists were rewarded for posting half-baked data on GenBank immediately upon discovery. The data would be fully baked in public, through comment and revision and linking, into something usable, available to all. Science broke into a sprint that has accelerated ever since. Old sequesterers who couldn't adapt retreated into tenured irrelevance.

Enter the next stage, again with Dan Janzen in the thick of it: the Barcode of Life. Janzen had been growling for years that all he wanted was a handheld device. "You find a bug, you rip a leg off it, you put it in the thing, the thing connects to the Internet, and it tells you what the bug is. If it doesn't know, it assumes you've found a new species and asks you details about it, and requests that you also insert a piece of the leaf you found the bug on." (That last bit shifts the subject from taxonomy to ecology, from identity to relationship.)

In 2003 Paul Hebert, at the University of Guelph in Ontario, developed the beginnings of the shortcut Janzen was seeking. Hebert discovered the diagnostic value of a mitochondrial gene that most animals have—because it is crucial for energy—and that is highly variable because it evolves quickly. The telltale gene fragment is only 648 base pairs long. That means it can be sequenced for $10 a specimen, and that changes the world.

By 2008 the DNA barcoders had analyzed 375,000 specimens. The earliest tests came from Dan Janzen and Winnie Hallwachs in Costa Rica, full of news. One species of skipper butterfly turned out to be 10 different species. As Janzen proceeded to barcode the 10,000 or so butterfly species in the Guanacaste, he discovered that "a standard result is that 20 morphologically-defined species turn into 60 barcode species! A spin-off is that all the generalists disappear: they turn out to be clusters of look-alike specialists." Brian Fisher at the California Academy of Sciences is barcoding all the ants in Madagascar. A project is under way to barcode the world's birds, and there's a campaign to get all the fish barcoded. Now that two diagnostic gene fragments for plant identification have been identified, barcoding of the world's plants is proceeding. The barcodes, as they emerge, are, of course, being posted on GenBank.

So what?

Well, *you* just acquired some very subversive power, similar to what you can do with a cellphone camera. Janzen's handheld DNA device isn't here quite yet, but in the meantime, you can ship bits of tissue to Guelph for a few bucks and learn amazing things, like whether you've been cheated at the fish market. Jesse Ausubel (the Green nuke lover) unleashed two teenage students on ten groceries and four restaurants in Manhattan, collecting fish samples for barcoding. According to the *New York Times:*

> They found that one-fourth of the fish samples with identifiable DNA were mislabeled. A piece of sushi sold as the luxury

treat white tuna turned out to be Mozambique tilapia, a much cheaper fish that is often raised by farming. Roe supposedly from flying fish was actually from smelt. Seven of nine samples that were called red snapper were mislabeled, and they turned out to be anything from Atlantic cod to Acadian redfish, an endangered species. . . . Two of the 4 restaurants and 6 of the 10 grocery stores had sold mislabeled fish.

Just as pocket calculators democratized math, DNA barcoding makes the whole world bioliterate. As Janzen says, "We can make it so that each of all 7 billion people can know what bit of biodiversity is biting them, appealing to them, worrying them, attracting them, itching them, sickening them, and providing whatever goods or services can come from being able to know what it is." Amateur bird-watchers transformed ornithology; now empowered with barcoding (and whatever follows it), amateur taxonomists of every stripe will transform our knowledge of life on Earth. All species may be identified. Ecology may become predictive after all.

✦

Ethnobotanist Gary Nabhan tells of an epiphany he had while comparing two maps. One showed the U.S. counties that had the most endangered species. The other displayed counties in terms of the length of people's residency. The pattern jumped out at him: "Where human populations had stayed in the same place for the greatest duration, fewer plants and animals had become endangered species; in parts of the country where massive in-migrations and exoduses were taking place, more had become endangered." Another correlation, explored in Nabhan's *Cultures of Habitat* (1997), is that, worldwide, regions of high natural diversity have high cultural diversity. Life is richest where culture is richest and most constant.

Can that be a goal, a strategy? The way to play Indian is not with feathers, but with attention. You don't have to be born in a place to be native to it, you just have to engage it long enough and deep enough to belong there. When that happens, you're well and truly home. The poet Gary

Snyder has dwelled since 1970 in a hand-built house in the west-facing foothills of the Sierra Nevada range in California. His e-mails end with his address:

Kitkitdizze

      north of the South Yuba River

          near the headwaters of Blind Shady Creek

    in the trees at the high end of a bunchgrass meadow

(Kitkitdizze is the Miwok Indian name for a pine understory shrub found around his place. Perhaps because it is sticky and sharp-smelling, its English name is mountain misery.)

Reinhabitation, we used to call it; also bioregionalism. In *CoEvolution* and elsewhere we published a reinhabitory quiz meant to put people on the spot about their natural-systems ignorance and inspire greater immersion in their locale. You can find a current version online under the title "The Big Here." All the versions begin with the injunction, "Point north." (Can you? Right now?) Then: What phase is the moon in? What local spring wildflower is consistently the first to bloom? Name five native edible plants nearby, and their best season. Name five local birds and say which are migratory. What indigenous tribe used to live where you live now? Is the soil under your feet more clay, sand, rock, or silt? (I fail that one.)

Water questions are particularly revealing because they link natural and artificial infrastructure: How far do you have to travel before you reach a different watershed? Can you draw the boundaries of yours? How deep do you have to drill before you reach water? Trace the water you drink from rainfall to your tap. When you flush, where do the solids go? And what happens to the waste water?

That leads to other built-infrastructure questions: Where is the nearest cellphone tower for your carrier? What does your regional power utility get its electrons from—coal, nuclear, natural gas, hydro, biofuels, solar, or wind? If you have natural gas in your home, where does it come from? Does your local utility have time-of-use charges that allow you to save money by using less electricity during peak hours? What is the surcharge for peak usage? What is the oldest still-occupied building in your town?

What parts of your local infrastructure are overdue for maintenance? (I don't know on that one, or on four of the others.)

Most important: How is climate change expected to affect your region? What are people doing about that?

The point is to build knowledge *about* where you live in order to better take responsibility *for* where you live. That's the "gardening" this chapter is about. I like Gary Snyder's meditation on the Buddhist version of the Conservation Pledge I grew up with:

> There is a verse chanted by Zen Buddhists called the "Four Great Vows." The first line goes: "Sentient beings are numberless, I vow to save them." *Shujo muhen seigando.* It's a bit daunting to announce this intention—aloud—to the universe daily. This vow stalked me for several years and finally pounced: I realized that I had vowed to let the sentient beings save *me*. In a similar way, the precept against taking life, against causing harm, doesn't stop in the negative. It is urging us to *give* life, to *undo* harm.

Snyder also helps make the link from earth discipline to whole-earth discipline. "Like it or not," he writes, "we are *all* finally 'inhabitory' on this one small blue-green planet. It's the only one with comfortable temperatures, good air and water, and a wealth of living beings for millions (or quadrillions) of miles."

There is harm to undo in this place. Earth as a whole is the most ambitious and necessary restoration project of all.

•┄┄┄┄┄•

Footnotes for this chapter may be found at **www.sbnotes.com**.

Planet Craft

·9·

After Sputnik, there is no nature, only art.

—Marshall McLuhan

Whether it's called managing the commons, natural-infrastructure maintenance, tending the wild, niche construction, ecosystem engineering, mega-gardening, or intentional Gaia, humanity is now stuck with a planet stewardship role. Paul Crutzen, the atmospheric chemist who won the Nobel Prize in 1995 for his work on ozone depletion, coined a word that has resonated. "It seems appropriate," he wrote, "to assign the term *Anthropocene* to the present, in many ways human-dominated, geological epoch." We are shaping Earth so profoundly that it is evident in the geological record. The atmospheric and ecological changes we have made are expected to reverberate for tens of thousands of years.

If humanity's role has expanded to the point that the entire Earth is our niche, the trend of the changes we have made lately indicates we are doing a poor job of niche maintenance. The signs can be seen in any current version of the photograph of the Earth from space that sparked the environmental movement. "That icon," Jim Lovelock wrote in *The Vanishing Face of Gaia*, "is undergoing subtle change as the white ice fades away, the green of forest and grassland fades into the dun of desert, and the oceans lose their blue-green hue and turn a purer, swimming-pool blue as they too become desert." Civilization needs Gaia more than Gaia needs civilization. With the growing deserts she is beginning to shrug us off. To head off that fate, we have to engage her processes at her scale, in a conciliatory mode.

We are forced to learn planet craft—in both senses of the word: craft as skill and craft as cunning. The forces in play in the Earth system are astronomically massive and unimaginably complex. Our participation has to be subtle and tentative, and then cumulative in a stabilizing direction. If we make the right moves at the right time, all may yet be well.

● One emergent principle might be that deleterious elements should be concentrated. Concentrating people in cities is good. Concentrating energy waste products like nuclear spent fuel in casks is an improvement over distributing the greenhouse gases from spent coal and oil in the atmosphere. Concentrating our sources of food and fiber into high-yield agriculture, tree plantations, and mariculture frees up more wildland and wild ocean to carry out their expert Gaian tasks.

A "natural infrastructure" approach to ecosystem services can be helpful if it doesn't try too hard to be economically rigorous. Price comparisons do help to inform some decisions. A UN analysis of the "total economic value" of cutting trees for a coastal shrimp farm in Thailand gave the advantage to undisturbed mangrove forest—worth $1,000 to $36,000 per hectare for its timber, charcoal, offshore fisheries, and storm protection— over the shrimp farm's $200 per hectare.

Some comparisons, though, have to be more qualitative than quantitative. Forest ecologist Herbert Bormann has written that once we have cut down a forest, "We must find replacements for wood products, build erosion control works, enlarge reservoirs, upgrade air pollution control technology, install flood control works, improve water purification plants, increase air conditioning, and provide new recreational facilities." That logic helps Daniel Janzen charge the Costa Rican government for ecosystem services provided by the Guanacaste Conservation Area, but it doesn't tell what price to set. That has to be negotiated.

The fact is that even our artificial infrastructure operates on fuzzy economics. Nearly all large-scale projects—bridges, dams, tunnels, railroads, airports, power plants, wind farms, transmission lines—come in way over budget and behind schedule, and they don't pay out as expected. One global study of sixty projects with an average cost of $1 billion "found that almost 40 percent of the projects performed very badly and were either abandoned totally or restructured after experiencing some kind of financial crisis." In another study of thirty-six such megaprojects, three quarters of them failed to meet financial expectations. The norm is: We make grand plans, we build stuff, we're mostly glad we did, and the money gets sorted out awkwardly over decades.

Excessively precise economic analysis can lead to assessing everything in terms of its easily measurable *melt value*—the value that thieves get from stealing copper wiring from isolated houses, that vandals got from tearing down Greek temples for the lead joints holding the marble blocks together, that shortsighted timber companies get from liquidating their forests. The standard to insist on is *live value*. What is something worth when it's working? We'll seldom get a precise valuation, but ballpark is OK.

For some infrastructure, though, there is no ballpark. What is a stable climate worth? What would we pay to keep the one we have? Is there some amount about which we would say, "Sorry, that's just too expensive. We have to let the climate go"? That calculation is not financial.

• For sensitive ecosystem engineering at planet scale, what we need most is better knowledge of how the Earth system works. We are model-rich and data-poor. We need to monitor in detail and map in detail what's really going on, and the measuring has to be sustained and consistent to get the all-important trends of change over time. As we learned with Gen-Bank, immediacy and transparency are crucial. Systems analyst Donella Meadows laid down the commandment: "Thou shalt not distort, delay, or sequester information. You can drive a system crazy by muddying its information streams. You can make a system work better with surprising ease if you can give it more timely, accurate, and complete information."

In particular, we need to cure our ignorance about the ocean. "We are running on theoretical vapor," Jim Lovelock wrote. "The ocean is truly *aqua incognita*. . . . It is right to build theories of the ocean even though we know so little about it, but quite wrong to use them to make policies. First they must be tested by long-term observation and measurement, and that I think should be our first priority." The air we breathe comes from the ocean; so does the rain; so do the clouds that regulate the Earth's albedo, which in turn governs the climate. Also, as oceanographer Sylvia Earle points out, the ocean "provides home for about 97 percent of life in the world, and maybe in the universe." That life, most of it microbial, determines most of the Gaian balance of gases in the atmosphere.

In 2009 the spectacular array of services from Google Earth was expanded to include Google Ocean. Besides displaying the best current data on the ocean bottom and on currents and temperature, it is adding Encyclopedia of Life material as it accumulates. Google Earth is being used to track the behavior of everything from polar ice to radio-tagged animals. Threatened habitat is monitored, and so are illegal logging and mining operations. In the United States, a Google Earth add-on called MapEcos flags all the industrial polluters, complete with detailed comparison with other offenders, and a service called Vulcan maps carbon dioxide emissions from fossil-fuel use.

Another vital project is a comprehensive Global Soil Map being assembled by the International Soil Reference and Information Centre "to help informed decisions not only about agriculture, but also to monitor the effects of climate change, environmental pollution and deforestation." The first stage is a detailed soil map of Africa. Because these maps live online in digital form, they will improve over time rather than becoming obsolete, as printed maps do. (I saw that happen with a *California Water Atlas* I instigated in 1979 while working for Governor Jerry Brown; the maps and diagrams in our book helped the state for only a few years.)

Tools for sophisticated sensing are proliferating. An Australian timber importer uses DNA analysis to track every log brought in from Indonesia, to be sure it was legally cut. (Some 80 percent of the wood coming out of Indonesia is said to be illegal.) Toxicity testing of the world's eighty thousand industrial chemicals is taking a leap forward with DNA chips replacing animal tests. Localized analysis of carbon dioxide flow is being measured in the United States by a service called CarbonTracker, by FLUXNET globally, and by a major project in India called IndoFlux. Whole regions of carbon dioxide and methane variations are being measured from space by Japan's Ibuki satellite, launched in 2009.

How do we make sense of what we measure? Blogger Cory Doctorow describes the growing flood of data as a "relentless march from kilo to mega to giga to tera to peta to exa to zetta to yotta." To be of use to science, the data must be correlated, calibrated, synchronized, and updated. *Wired* observed that "Earth is peppered with high tech monitoring hardware from polar-orbiting satellites to instrument-laden buoys. Problem is, they're all operating in Babel-style disconnect." Efforts are under way to link everything in a mutually intelligible way via a Global Earth Observation System of Systems, and what are called Data-Intensive Scalable Computer systems are expanding search capabilities.

Many of the databases welcome amateur input. For instance, reports from gardeners and students on the changing nature of seasonal phenomena (phenology) are being collected at BudWatch in the United States, NatureWatch in Canada, Nature's Calendar in the United Kingdom, and De Natuurkalender in Holland. When do the lilacs bloom where you are? When do the first swallows show up?

Piece by piece, we are building a digital Gaia.

● The biggest Earth-monitoring story of the twenty-first century's first decade is one that didn't happen. It began in 1998, when Vice President Al Gore hatched an idea for a space camera that would provide a constant real-time, high-resolution video of the Earth turning in the sunlight, both for inspiration and for science. Its location would be at the Lagrange-1 point of neutral gravity between Earth and the Sun, a little under a million miles from here. As seen from L-1, the Earth's disk would always be fully lit—the camera would have the Sun perpetually behind it. That location in space had already proven its value with the Advanced Composition Explorer satellite, launched by the United States in 1997. The ACE monitors the Sun, giving an hour's advance warning of "geomagnetic storms that can overload power grids, disrupt communications on Earth, and present a hazard to astronauts," according to the mission's Web site.

The Republican-dominated Congress ridiculed Gore's idea ("Al's screensaver") and sent it to the National Academy of Sciences for review and presumed disposal. Instead, the scientists urged that the project go ahead with a package of sophisticated scientific instruments on board for observing Earth. Named the Deep Space Climate Observatory (DSCOVR), it would measure variations in Earth's ozone levels, aerosols, water vapor, cloud thickness, and the reflected and emitted radiation—the total energy budget—of the whole planet. "DSCOVR would offer a global, rather than myopic, perspective of the planet," said the mission's principal investigator, Francisco Valero, of Scripps Institution of Oceanography. (I can imagine Jacques Cousteau cheering.) It would also calibrate the instrument readings from all the low-Earth-orbit satellites that we currently rely on.

Still Republican-dominated, Congress approved the money—$100 million—and the satellite was built, ready for launch in 2001. But between the construction and the launch there was an election. The incoming Bush administration was hostile to Gore, to science in general, and to climate science in particular. Out of spite, the new administration postponed and then canceled DSCOVR's launch. France and Ukraine both offered to launch the satellite for free, and both were turned down. In 2008 a gathering of forty-four leading climatologists in Germany declared that an Earth observation satellite at L-1 is "essential" for climate science.

When I wrote this in early 2009, I could only hope that the Obama administration would—as it promised Al Gore—finally launch DSCOVR (the satellite was quietly defended from the dismantlers), and that its data and grand imagery would soon be giving us Earth live. That would be a happy ending to the story. But whatever happens, we lost nine years of critical data to partisan folly. Politics trumped science.

An editorial in *Nature* said, "There is only one Earth, with only one history, and we get only one chance to record it. . . . A record not made is gone for good."

✦

The idea of adjusting Earth's climate directly is anathema to most, for good reason. All geoengineering schemes look like attempts to sow the wind that are sure to reap the whirlwind. The dangers are certain, enormous, and inescapable. The imagined benefits rely on chaotic mechanisms and unproven theory. Surrendering to such shortcuts is the height of irresponsibility, we tell each other. But the tenor of the discussion is changing, and geoengineering is being taken seriously, sooner than expected, because of emerging realizations.

Realization 1. The stupendous cost, disruption, and time required to build a low-carbon energy infrastructure—Saul Griffith's Renewistan—is sinking in. We will contemplate the price, over twenty-five years, of 30,000 square miles of solar electric cells, 15,000 square miles of solar thermal collectors, 1.5 million square miles of algae farms for biofuel, 2.6 million wind turbines and the space they take up (about 100,000 square miles), 27,400 geothermal steam turbines, and 3,900 1-gigawatt nuclear reactors— not to mention the cost and disruption of shutting down the coal, oil, and gas infrastructure that all that Green technology is supposed to replace, nor the environmental burden of covering the natural landscape with a continent's worth of hardware.

(The day I wrote that paragraph I got a bulletin from the California Native Plant Society urging its nine thousand members to protest against eighty planned solar-energy projects that will cover 1,000 square miles of wild desert in the southern part of the state: "The potential impact of these projects to rare plants, vegetation, animals, and intact, majestic desert landscapes is unfathomable. . . . In the construction process, solar energy developments effectively denude the landscape and leave little intact habitat.")

Realization 2. It will become painfully apparent that mitigation is not going to succeed. The whirlwind is coming anyway. Currently imaginable efforts to reduce greenhouse gas emissions do not level off at the desired 450 parts per million (ppm) of CO_2 in the atmosphere, nor at 550 ppm, and probably not even at 650 ppm. Increasingly vivid knowledge of how lethal a 650-ppm world would be will motivate a frantic search for alternative paths.

Realization 3. Minds change with events, though usually it takes several in succession. The war in Darfur has not been seen as the drought-driven resource crisis it is. The death of 35,000 in Europe's heat wave of 2003 was considered an anomaly rather than a window on the future. But more such events will pile up. Cyclone Nargis, which hit Burma in May 2008 and killed over 150,000, was the seventh-deadliest cyclone of all time; another climate-derived cyclone making landfall slightly farther west would simply drown Bangladesh. As the Tibetan Plateau dries up, the reduced flow in the many rivers it feeds will set downstream nations at war with those upstream. Vietnam, Cambodia, Thailand, Laos, and Burma could ally to fight China (or one another) over control of the dwingling water in the Mekong River. Nuclear-armed India might cut off the flow of the Indus River into nuclear-armed Pakistan, which would surely bomb in response. Climate change will kill some people directly, but most will die at the hands of other people made desperate by climate change. When that happens, there will be demand for action on climate that shows immediate results.

Realization 4. News from field climatologists will keep getting worse. When one positive feedback—such as a "gigaburp" of methane released from melting permafrost—takes off conspicuously, a sense of public emergency will take off with it. Already temperatures in the Arctic have gone up over 4°C since 1950. The suddenness of a self-accelerating phenomenon invites proportionally immediate response.

Realization 5. Some forms of geoengineering, expensive as they are, may be a hundred to a thousand times cheaper than building Renewistan, and some of them would have an instantaneous effect on climate rather than one delayed by decades. As soon as climatic conditions become frightening and urgent, geoengineering schemes will suddenly jump from "plausible but dangerous" to "dangerous but we have no choice." The cost is low enough that a single nation or even a wealthy individual could set in motion a geoengineering project that would affect everyone on Earth. (A growing number of workshops are addressing the specter of unilateral geoengineering.)

• Any one of those realizations is sufficient; in combination they are overwhelming. Geoengineering schemes will be in high demand shortly, but what exactly is on offer? Here's the catalog as of 2009, in order of the likelihood of their being attempted, with the most tempting first: global dimming with stratospheric sulfates, brightening the Earth with clouds from ocean spray, feeding iron to ocean phytoplankton to increase their fixation of carbon, floating vertical pipes in the ocean for the same purpose, converting agricultural waste into biochar, massive air capture of atmospheric carbon, and global dimming with mirrors in space. No doubt more ideas will emerge—and should emerge; that's a paltry list, considering the potential need. But examining these will give a sense of the ingenuity, daunting scale, and potential hazards of geoengineering strategies.

Employing stratospheric sulfates is the first choice of most climatologists because it has already been proven to work. In 1991 a volcano in the Philippines, Mount Pinatubo, erupted explosively, sending 20 million tons of sulfur dioxide twenty miles up into the stratosphere, where the material oxidized into tiny sulfate droplets that absorbed and reflected sunlight. The following year, the entire planet cooled by half a degree Celsius. Sea ice in the Arctic was so durable that the crop of particularly large and healthy young polar bears born in 1992 were called the Pinatubo cubs.

In 1998, at a climate conference in Aspen, a space weapons expert (and microreactor designer) named Lowell Wood gave a provocative presentation on the stratospheric-sulfates scheme. Climate modeler Ken Caldeira was so annoyed that he set out to prove the idea couldn't work; instead his models suggested it might work very well, with relatively few side effects. Caldeira became a convert to geoengineering and began cowriting papers with Wood. In 2006 Paul Crutzen published an essay in *Climatic Change* that signaled a major shift in scientific opinion. He found international efforts to reduce carbon dioxide emissions so "grossly disappointing" that a backup plan such as "albedo enhancement by stratospheric sulfur injections" must be explored. He wrote:

> If sizeable reductions in greenhouse gas emissions will not happen and temperatures rise rapidly, then climatic engineering, such as presented here, is the only option available to rapidly reduce temperature rises and counteract other climatic effects. Such a modification could also be stopped on short

notice, if undesirable and unforeseen side effects become apparent, which would allow the atmosphere to return to its prior state within a few years. . . . Provided the technology to carry out the stratospheric injection experiment is in place, as an escape route against strongly increasing temperatures, the albedo adjustment scheme can become effective at rather short notice, for instance if climate heats up by more than 2°C globally or when the rates of temperatures increase by more than 0.2°C/decade.

Proposed methods of delivering the sulfur to the stratosphere include airplanes, cannons, and balloon-suspended hoses. One or two million tons of sulfur a year could keep the Earth's temperature level even if greenhouse-gas emissions doubled. As Ken Caldeira put it, "If we could pour a five-gallon bucket's worth of sulfate particles per second into the stratosphere, it might be enough to keep the Earth from warming for fifty years. Tossing twice as much up there could protect us into the next century." For perspective, remember that humanity at present releases yearly the equivalent of five Pinatubos—100 million tons of sulfur dioxide pollution—into the lower atmosphere, where its dimming effect keeps the Earth 2°–3°C cooler than if our air were clean. The estimated cost of injecting stratospheric sulfur would be $1 billion a year, which is shockingly little, considering its impact.

The way to test the technique, Caldeira and his colleagues propose, is with a localized effort in the Arctic. A relatively small amount of sulfur could be used; it could be injected into the lower stratosphere, so it would stay up only a year; few people live in the region; and the area needs cooling more than any other. The effects could be measured directly in ice behavior (and polar bear cubs), and increased ice would amplify the cooling effect by reflecting even more sunlight. It could be, editor Oliver Morton wrote in *Nature*, "as low-impact an option as the geoengineer's toolbox offers."

● Even more attractive, in terms of the ability to turn it off easily, is the idea of a fleet of oceangoing cloud machines. In 1990 atmospheric physicist John Latham came up with the idea of significantly brightening Earth's albedo by simply adding more water droplets to the stratocumulus clouds that cover a third of the oceans. The droplets could come from atomized seawater. Engineer Stephen Salter designed a ship to do the job. It would be an unmanned pontoon ship, 150 feet long, controlled by satellite, utilizing towerlike Flettner sails to course back and forth across the wind. (The vessel is incredibly cool looking.) Turbines dragged beneath the surface would provide the power to spray seawater in 1-micron droplets at a rate of 500 gallons a minute. Latham and Salter estimate that 1,500 such ships, which would cost a total of about $3 billion to build, would brighten ocean clouds enough to offset a doubling of carbon dioxide in the atmosphere. Lovelock comments: "Because this approach has far fewer potential adverse side effects than the stratospheric aerosols, it should be tried on a sufficient scale to assess its worth."

• So far, the most controversial geoengineering proposal has to do with feeding iron to the ocean's carbon-fixing algae (also called phytoplankton, diatoms, and coccolithophores). Vast regions of the ocean surface are virtually devoid of phytoplankton, and no one could figure out why until biochemist John Martin suggested in 1990 that what makes the difference is the presence or absence of iron in dust blown from the land. A dozen experiments have proved his hypothesis correct: fertilizing the ocean with iron makes huge algal blooms. Now the question is, does the extra carbon fixed by the algae sink into the ocean depths, or does it circulate in food webs near the surface and return to the atmosphere as carbon dioxide? If the carbon sinks in the form of dead algae below a depth of 1,600 feet, then it will stay out of the atmosphere for at least a century; if it gets all the way to the bottom, it could be gone for thousands of years. During past ice ages, when dry land sent more iron-bearing dust into the oceans, 100 billion tons of carbon were sequestered in the sea.

In 2004 German oceanographer Victor Smetacek led an expedition of fifty scientists to conduct an iron-fertilization experiment in the ocean between South Africa and Antarctica. One month after the team dumped 3 tons of iron filings in the water, they detected large quantities of dead algae sinking many hundreds of feet below the extensive algal bloom, but they lacked equipment for detailed research at depth. In early 2009, Smetacek led another team (this time including many scientists from India's National Institute of Oceanography) to the Southern Ocean with a plan to study the deep-ocean effects of fertilizing algal blooms with 20 tons of iron. *Science* reported:

> The new experiments will explore what happens to those blooms and whether they can be carbon sinks for atmospheric carbon dioxide. There's a lot scientists don't know, including why some blooms fall so rapidly, how much of them are devoured by microbes and other sea life on the way down, and which locations and plankton species do the best job of sequestering carbon.

ETC, the anti-genetic-engineering group in Canada, raised a howl of protest, quoting an environmental lawyer from South Africa's Center for Biosafety: "We do not believe our country should be aiding and abetting these controversial geoengineers in breaking the global moratorium. We have formally asked our Environment Ministry to compel the ship to return to port and offload its cargo of iron." The moratorium referred to is a 2008 agreement (pushed by ETC and others) within the UN Convention on Biodiversity to cease ocean fertilization activities until there is "an adequate scientific basis on which to justify such activities." In other words, in seeking to block the Smetacek research expedition (which was approved by the German and Indian governments), ETC and other environmental organizations were saying, "You must have scientific proof of the effectiveness of iron fertilization, and we will prevent you from getting it."

The Bush administration blocked research on climate change for identical reasons—fear of findings that might go against an ideological position. Neither climate denialists nor ETC seem to consider that the research could go in their favor. They would rather believe than know.

There is no excuse for environmentalists to block scientific research on environmental issues, ever.

• Jim Lovelock and Chris Rapley (former British Antarctic Survey director; now director of London's Science Museum) have put forth an idea that is less controversial than iron fertilization but similar in purpose: An array of floating vertical pipes would provide nutrients to organisms near the ocean's surface by drawing cold, nutrient-rich water up from the depths. The point is to break through the thermocline in stratified waters and deliver the nutrients to where the sunlight is, mimicking in miniature the upwellings that provide the ocean's areas of greatest productivity and biodiversity. Lovelock and Rapley wrote in *Nature:* "Water pumped up pipes—say, 100 to 200 metres long, 10 metres in diameter and with a one-way flap valve at the lower end for pumping by wave movement—would fertilize algae in the surface waters and encourage them to bloom. This would pump down carbon dioxide and produce dimethyl sulphide, the precursor of nuclei that form sunlight-reflecting clouds." With three-foot surface waves, the pipes would send four tons of cool water a second to the surface, using the ocean's energy to do the work. Besides enriching biodiversity, the localized cooler water could protect coral reefs endangered by ocean heating, and weaken cyclones in places like the Gulf of Mexico by depriving them of the overheated seawater that turns a category 3 hurricane into a category 5 Katrina.

Geophysical Research Letters reported in 2008 that, over nine years, global warming has caused a 15 percent increase in barren regions of the ocean, due to the warmer water stratifying. Fewer clouds form over dead ocean regions for lack of the cloud-forming droplet nuclei usually provided by life on the ocean surface. If you take a world map of ocean chlorophyll and overlay it with a map of ocean cloud cover, you find watery deserts of enormous size—such as a region of the Pacific 1,000 miles wide stretching west from Latin America for 5,000 miles along the Tropic of Capricorn. In these ocean deserts, there is a dearth of life and clouds and an overabundance of sunlight heating the dark water and spreading the deadly stratification. It's a double-whammy positive feedback—like the melting of polar ice and the drying of rainforests combined: Ever more sunlight is being absorbed *and* there is ever less plant life to fix carbon.

● Then there's biochar. It looks like such a miracle material that I'm skeptical about its ability to scale up. Jim Lovelock is convinced that it can do so. He cites a paper by Cornell soil scientist Johannes Lehmann in *Nature* proposing that biochar is a more reliable method of sequestering carbon than just planting trees (they can burn), converting to no-till agriculture (the carbon-fixation gains level off in twenty years), or using geological storage sites for carbon dioxide (they can leak). "Once biochar is incorporated into soil," Lehmann wrote, "it is difficult to imagine any incident or change in practice that would cause a sudden loss of stored carbon."

In 2009 Lovelock told a *New Scientist* interviewer what it would take for biochar to scale up and what it would mean:

> There is one way we could save ourselves and that is through the massive burial of charcoal. It would mean farmers turning all their agricultural waste—which contains carbon that the plants have spent the summer sequestering—into non-biodegradable charcoal, and burying it in the soil. Then you can start shifting really hefty quantities of carbon out of the system and pull the CO_2 down quite fast. . . . Ninety-nine percent of the carbon that is fixed by plants is released back into the atmosphere within a year or so by consumers like bacteria, nematodes and worms. What we can do is cheat those consumers by getting farmers to burn their crop waste at very low oxygen levels to turn it into charcoal, which the farmer then ploughs into the field. A little CO_2 is released but the bulk of it gets converted to carbon. You get a few percent of biofuel as a by-product of the combustion process, which the farmer can sell. This scheme would need no subsidy: the farmer would make a profit.

The world's volume of timber and crop residue that might be pyrolyzed into biochar is truly massive. How sweet it will be if the *terra preta* technique invented thousands of years ago by Amazon Indians to solve their soil problems works today to help solve our atmosphere problems.

• A synthetic method for fixing carbon directly from the air is being pursued by an environmental engineer, Klaus Lackner at Columbia University. He proposes "artificial trees" that would capture carbon a thousand times more efficiently than living trees of the same size. Allen and Burt Wright at Global Research Technologies, based in Tucson, Arizona, have developed a working prototype. The device flows sodium carbonate down sheets of a proprietary plastic. The liquid reacts with carbon dioxide to form baking soda. Electrolysis separates the carbon dioxide from the baking soda and cycles the sodium carbonate for reuse. An extractor the size of a shipping container, Allen Wright claims, could capture a ton a day of CO_2 at a cost of $30 a ton. The energy to run the extractor would have to come from a non-fossil-fuel source for the process to come out ahead on carbon, however.

What do you do with captured carbon dioxide? Because it is an industrial chemical (used in greenhouse horticulture, oil field enhancement, food processing and transport, water treatment, foam fabrication, dry ice, etc.), the company proposes collecting the CO_2 at the site of use for commercial sale. There are also thoughts of "mineral sequestration"—forcing carbon dioxide, with heat or acid, to react with serpentine or peridotite rock to form extremely stable magnesium carbonate. "This is what nature eventually does anyway," says Lackner. "Our goal is to take a process that takes 100,000 years and compress it into 30 minutes."

• I first heard the idea of putting sunglasses directly on the Sun from Jim Lovelock in 1986; it scandalized and thrilled me at the time. These days it's a serious proposal. The magical spot is the one Al Gore chose for the Earth-imaging and -sensing satellite DSCOVR—the Lagrange-1 site of neutral gravity between Earth and the Sun. (Gore hates the idea of solar shades at L-1. When I mentioned it to him, he said, "Oh right, Brand. Let's just experiment with the whole planet!")

There are several schemes afoot. The one that gets the most discussion so far comes from Roger Angel, a University of Arizona astronomer renowned for his work on telescope mirrors. His proposal, titled "Feasibility of Cooling the Earth with a Cloud of Small Spacecraft Near the Inner Lagrange Point (L1)," appeared in the *Proceedings of the National Academy of Sciences* in 2006. To dissipate 1.8 percent of the sunlight reaching Earth, which would be enough to offset a doubling of CO_2 in the atmosphere, Angel would float 16 trillion two-foot disks in a cloud eight thousand miles wide and sixty thousand miles long, aligned between the Sun and Earth. The disks, each weighing a gram, could keep themselves on station "by modulating solar radiation pressure, with no need for expendable propellants." (That means that the disks tack in the "wind" of sunlight like sailboats.)

To get 20 million tons of disks to L-1, Angel suggests electromagnetic rail guns for launch and ion-drive rockets for flight. *New Scientist* reported: "Angel has calculated that 20 rail guns, each 3 kilometers high, working round the clock and launching one bundle of discs every 5 minutes for 10 years, could put enough 'pico-satellite' discs into space to provide the required 1.8 percent reduction in the sunlight reaching Earth. The cost would be in the region of a few trillion dollars." (Angel ended his *PNAS* paper with these words: "The same massive level of technology innovation and financial investment needed for the sunshade could, if also applied to renewable energy, surely yield better and permanent solutions.")

✦

Heady stuff. One's immediate response ricochets from "How dangerous and crazy!" to "How grand and thrilling!" to "How handy and cheap!" and back to "How dangerous and crazy!" But what happens when the considered response is "How necessary"? In late 2008, an English newspaper, the *Independent*, polled eighty climatologists about geoengineering. Over half of them said that the situation is growing so dire that we must have a geoengineering backup plan, while 35 percent said that such plans would distract people from the crucial task of reducing greenhouse gases (11 percent offered no opinion).

The first need is to run some large-scale experiments. The projects have to move from conceptual design to operational design—from scientists to engineers—with serious money attached. "There's a lot more talk than work," Ken Caldeira told *Scientific American*. "Most of the research has been at the hobby level." Environmental scientists Thomas Homer-Dixon and David Keith wrote in the *New York Times* that geoengineering "is so taboo that governments have provided virtually no research money." They urged that we should begin with "real-world tests of various technologies that poke the climate system just a little." They concluded: "The important thing is to get scientists, environmentalists and global-warming skeptics alike out of the nonsensical all-or-nothing dichotomy that characterizes much current thinking about geoengineering—that we either do it full scale, or we don't do it at all."

So test the ideas we have, and keep thinking up new ones. For example, there's a scheme to ionize carbon dioxide molecules with lasers, which would supposedly cause the Earth's magnetic field to eject the CO_2 from the atmosphere at the poles. I would outline the concept here if I understood it, but I don't. Far more comprehensible is an idea, proposed in 2009, called CROPS—Crop Residue Oceanic Permanent Sequestration. Taking 30 percent of U.S. agricultural residues and sinking the baled material in the ocean would capture and sequester 15 percent of the annual increase of global atmospheric CO_2, according to bioremediation expert Stuart Strand and physicist Gregory Benford. I like the criteria they use to judge a sequestration scheme's practicality. As summarized by *Science Daily*, the method must "move hundreds of megatons of carbon, sequester that carbon for thousands of years, be repeatable for centuries, be something that can be implemented immediately using methods already at hand, not cause unacceptable environmental damage, and be economical."

● How does the current roster of notions stand up to criticism? Oliver Morton reported in *Nature*: "Geoengineering, many say, is a way to feed society's addiction to fossil fuels. 'It's like a junkie figuring out new ways of stealing from his children,' says Meinrat Andreae, an atmospheric scientist at the Max Planck Institute for Chemistry in Mainz, Germany." Quoting MIT climatologist Ronald Prinn—"How can you engineer a system whose behavior you don't understand?"—Morton answered the question: "As carefully and reversibly as you can."

There are two major worries about the stratospheric-sulfates approach. (They apply also to oceanic cloud brightening and sunshades in space.) Effectively dimming the Sun would relieve the pressure to reduce carbon dioxide in the atmosphere, but the short-term gain has a long-term penalty. Humanity would be forced to keep on dimming the Sun indefinitely, because if the project ever stopped, the renewed full sunlight on an atmosphere further overloaded with CO_2 would make the global temperature jump catastrophically. One climatologist calls it Damocles world.

Furthermore, the acidity of seawater would continue to increase, along with whatever harm that does to the ocean's vital ability to fix carbon. But there may be a ray of hope on that subject. Débora Iglesias-Rodríguez at the National Oceanography Centre in Southampton, UK, has found that increasing acidity is *welcomed* by one of the most abundant of all phytoplankton, a coccolithophore named *Emiliania huxleyi*—Ehux to its many friends (it has its own home page). Ehux not only thrives in acidic water, its rate of calcification goes up proportionally as acidity increases. The more carbon dioxide in the atmosphere and the ocean, the more Ehux fixes carbon in a form that sinks. In addition, algal blooms of Ehux—they are visible from space—increase the ocean's reflectivity. The blooms themselves are light-colored, and—in one of the first Gaian mechanisms to be identified—they emit dimethyl sulfide particles that lead to local cloud formation. Thus Ehux appears to work in three ways to head off climate change. It is a genuine natural *negative* feedback process that moderates global warming. As the Earth heats up, geoengineer Ehux responds by fixing more carbon and brightening the ocean's albedo, which helps Earth cool. Get this organism a grant. (Once the ocean surface stratifies, though, Ehux starves along with everything else.)

The carbon fixation schemes—biochar, ocean fertilization, air capture—have the double advantage of threatening little harm and working toward the permanent solution of getting to a manageable level of CO_2 in the atmosphere. But they're too slow. The same is true of just cutting back on carbon dioxide emissions, nearly impossible as that is. Lovelock points out: "The response time of the Earth to carbon dioxide change is of the order of 100 years."

To turn down the heat we will probably have to do something radical to alter what the Earth does with sunlight—dim it or reflect it.

● There remains a larger issue. Suppose geoengineering works technically. How can it work politically? Who decides to do it? Who runs it and balances its various forms? Who pays for it? Who accepts responsibility? Who compensates those who are harmed by it? Who decides which claims for compensation are valid? Does taking responsibility for climate geoengineering also mean taking responsibility for climate refugees?

It's easy to govern a negative. No one has to take responsibility for global warming because it was brought about by damn near everyone, and unintentionally. But geoengineering is intentional, an act of commission. Everything it accomplishes and fails to accomplish and inadvertently causes and is accused of causing has an identifiable human source. The climate will keep changing as it is without governance. To change the climate—the world—in the direction we want requires forms of governance we do not yet have.

Someone who has given the subject realistic forethought is David Victor, a Stanford law professor specializing in climate-change issues. Here I'll draw mainly on a 2008 paper he wrote for the *Oxford Review of Economic Policy* titled "On the Regulation of Geoengineering." People imagine, Victor says, that what is needed is a "legally binding regulatory treaty" like the Montreal Protocol governing harm to the ozone layer. But he thinks that "most treaties on geoengineering will be useless or actively harmful because, at present, experts and governments do not know enough about the scope and hazards of possible geoengineering activities to frame a meaningful treaty negotiation." He especially worries about treaties that would make geoengineering taboo, because "a taboo is likely to be most constraining on the countries (and their subjects) who are likely to do the most responsible testing, assessment, and (if needed) deployment of geoengineering systems. A taboo would leave less responsible governments and individuals—those most prone to ignore or avoid inconvenient international norms—to control the technology's fate."

Because the cost of some geoengineering schemes is so low, Victor predicts, "A lone Greenfinger, self-appointed protector of the planet and working with a small fraction of the Gates bank account, could force a lot of geoengineering on his own." The way to head off unilateral geoengineering and premature treaties, Victor suggests, is with a growing body of *norms* rather than rules:

> Meaningful norms are not crafted from thin air. They can have effect if they make sense to pivotal players and then they become socialized through practice. . . . Useful norms could arise through an intensive process of research and assessment that is probably best organized by the academies of sciences in the few countries with the potential to geoengineer. . . .
>
> Most likely . . . is that the impacts of global climate change will have reached such a nasty state by the time societies deploy large-scale geoengineering that some side effects will be tolerated. The . . . systems they deploy will not be a silver bullet but rather many interventions deployed in tandem—one to focus on the central disease and others to fix the ancillary harms.

To my mind, a useful role for Greenfinger entrepreneurs might be to jump-start serious geoengineering research while national academies of science are spending years making up their minds to act. Then the privately funded researchers could bring real data to the "transnational assessment process," where the norms and best practices emerge. This is a planetary hack we're talking about. It has to be totally transparent and highly collaborative. Everyone's first preference is to not deploy it at all, but if it has to be used, it must be done effectively and minimally, and if possible, for a limited period. Like abortion, geoengineering should be "safe, legal, and rare."

That still leaves the question of who runs things—"whose hands will be allowed on the thermostat," as David Victor puts it. The task can be divided between the operators and an oversight body. In one previous piece of planet craft—the total eradication of smallpox in the 1970s—the World Health Organization provided oversight and funding, and the Smallpox Eradication Unit, led by Donald Henderson, did the work.

In Victor's formulation, norms and leadership for geoengineering will emerge from an intensifying sequence of conferences, research projects, data sharing, and brainstorming. The most effective early players will determine the play, and funding will determine the pace. Geoengineering is government-scale infrastructure; it will need government-scale money. Once one nation commits, I suspect, other nations will join in, lest they be left out. If China says, "We're going to geoengineer," the United States, Russia, the European Union, Japan, Brazil, and India are not going to say, "Fine, let us know how it works out." They'll start their own programs. With luck, an ad hoc standards-setting body similar to the Internet Engineering Task Force ("rough consensus and running code") will emerge. That kind of governance was required in order to have one universal Internet. The planet's one universal climate requires something similar.

• We can get practice on how to engineer the Earth ecosystem with a light touch by stepping up to a bit of solar system engineering first. Earthlings now have the ability, but not yet the will, to prevent devastating asteroid and comet strikes. Astronaut Rusty Schweickart has led the way on this one. (He's had occasion to take orbital mechanics more seriously than most. On the Apollo 9 mission in 1969, which took place entirely in Earth orbit, he flew the Lunar Module 111 miles away from the Command Module. Because his craft could not reenter the atmosphere without burning up, he had to navigate back to the Command Module and dock with it, or die.)

Schweickart estimates that the probability of an "unacceptable" asteroid collision in this century is about 20 percent. Over the long term, of course, it's 100 percent. The impact from an asteroid over a kilometer in diameter would kill billions of people and violently disrupt the climate and biosphere. As of late 2008, NASA had detected 742 near-Earth objects (NEOs) of that size. Powerful new telescopes in Hawaii and Chile (the one in Chile partially funded by Greenfinger billionaires Charles Simonyi and Bill Gates) are expected to detect 21,000 near-Earth asteroids greater than 460 feet in diameter—a size that could wreck a nation or a seacoast—and up to 400,000 that could cause substantial harm. (The exploding asteroid that laid waste to 800 square miles of the Tunguska region in Siberia in 1908 was only 160 feet in diameter.) Very soon, dozens of asteroids will be identified as direct threats to Earth, requiring some kind of action.

How do you deflect an asteroid? Schweickart spells out the current technique:

> If you know for certain that an impact is due, it's generally too late to do anything about it. You have to act at the stage when it's a question of probabilities. Deflection of a potentially threatening asteroid is a relatively inexpensive three-step process, which you need to begin 15–20 years before the possible impact of the asteroid:
>
> 1) Place a transponder by any NEO that looks like it's big enough and in an orbit threatening to impact Earth. The transponder is necessary to get the detailed tracking information to decide what to do next. *If* the data indicates a collision is possible (like a 1-in-20 chance), proceed to step 2:
>
> 2) The first deflection move is kinetic—you "rear-end" the asteroid with a spaceship. The impact will drive it forward just enough ($^5/_{1,000}$th of a mile per hour or so) to miss Earth. But that still leaves the possibility that the asteroid might pass through one of several "keyholes" that would bring it back on a collision course with Earth in a later orbit. To adjust for that, go to step 3:
>
> 3) Use a "gravity tractor" to fine tune the orbit so that it misses the keyholes as well as the Earth. Adjustments are on the order of 1 to 10 millionths of a mile per hour. That should assure a relatively permanent solution for that particular rock.

(Because asteroids tumble, you can't just push them. In 2005 astronauts Edward Lu and Stanley Love devised the "gravity tractor": A spacecraft hovers gravitationally close to an asteroid and drives gently in the desired direction, drawing the asteroid with it.)

Schweickart says that asteroid deflection is similar to geoengineering for climate change, "only much simpler, better understood, and cheaper." As with climate, you're taking global action on a global problem, based on global awareness. "But the fatal missing element," he says, "is there is no agency in the world charged with protecting the Earth against NEO impacts." To remedy that, Schweickart got the Association of Space Explorers (an organization of astronauts and cosmonauts he founded) to send a "Draft NEO deflection protocol" in 2009 to the UN Committee on the Peaceful Uses of Outer Space.

The draft protocol recommends that the UN establish a three-part decision apparatus—one part to manage asteroid data, one to design missions, and one for oversight. Execution would be done by one or more of the major spacefaring entities—Russia, the United States, the European Space Agency, Japan, China, the United Kingdom, and India. To prove the overall concept and advance the process, Schweickart and others are lobbying the U.S. Congress and NASA to dedicate an American space mission "to significantly alter the orbit of an asteroid, in a controlled manner, by 2015."

As usual with grand projects, it's important to ponder the victory condition. What happens when you succeed? "There is one long-term consequence," Schweickart notes. "The Earth has been bombarded with rocks for 4.5 billion years. They made life and then shaped it. We'll be putting an end to that process. No more cosmic pruning of the tree of life. No more craters."

Asteroids can be carrot as well as stick, claims John Lewis in *Rain of Fire and Ice* (1996). Once we're able to redirect some, others we may wish to mine. "The *smallest* known metallic (M-type) asteroid, 3554 Amun," Lewis writes, has "a radius of 500 meters. It contains over $1,000 billion worth of cobalt, $1,000 billion worth of nickel, $800 billion worth of iron, and $700 billion worth of platinum metals. . . . By comparison, the uncontrolled impact of Amun with Earth would deliver a devastating 80,000-megaton blow to the biosphere, killing billions and doing hundreds of trillions of dollars worth of damage." (I should quickly add that asteroids of any size are worthless as weapons: They're hell to aim. Climate is no good as a weapon either, for the same reason.) If we start mining asteroids, space-based solar electricity for Earth would be an obvious use for the metals.

Asteroid deflection is such a crisp and doable program of planet craft, it can serve as a model and inspiration for the much more complex task of climate restoration.

✦

Seize the century. We're facing multidecade, multigeneration problems and solutions. Accomplishing what is needed will take diligence and patience—a sustained *bearing down*, over human lifetimes, to bridge the long lag times and lead times in climatic, biological, and social dynamics, and to work through the long series of iterations necessary for any apparent solution to become practical. At the same time, we need a professional caregiver's sense of urgency. Here's how that works.

You're outside the locked door of a room where you think a suicide might be going on. What do you do?

If you break down the door, it might be a false alarm. It could be that the person isn't there at all, or is just sleeping, and they'll be really upset about your breaking their door and embarrassing them. It might even make their emotional fragility worse. Besides, breaking down a door is hard. You might fail or hurt yourself. Other people in the building might try to stop you. The liability issues are horrible to contemplate.

Suppose the person in there really is committing suicide. It's his choice. What right do you have to interfere? If you do intervene, the person may well just try again later, and all you will have done is extend their suffering and delay the inevitable. On the other hand, suicide attempts are often impulsive and situational, caused by a medication imbalance or a wave of bad news. This may be a unique situation, a survivable crisis.

In life-threatening situations, time is of the essence. Getting to a stricken person fast often makes the difference between life and death or between life and permanent impairment. You don't have time to argue with yourself or with others at the door.

You're outside the locked door of a room where you think a suicide might be going on. What do you do? You break down the door.

● There is so much work to do that it doesn't matter who does it. Large corporations making money doing the right thing is just fine. The United Nations sending black helicopters to do the right thing is just fine. Property-defending conservatives doing the right thing is just fine. Placard-waving leftists stopping the wrong thing is just fine. Paul Hawken's myriad micro-organizations doing the right thing locally is the health of a system curing itself.

I would love to see the environmental movement as a whole (along with everyone else) embrace the kind of ideas posed here, and maybe it will happen. But my impression is that movements don't change that way. The environmental movement became unified in 1970 with Earth Day, and that unity served it well for a decade or two. Then the advantages diminished and the papered-over unity became more of a problem than a help. I would not be surprised if the movement now divides like a bacterium into two or more lively offspring.

In that case, there will be traditional Greens, ever more resolute in their well-established methods and purpose, and there will be another set of Greenish players more interested in innovation and risky endeavors. The new crowd will eventually be labeled—Post-Greens, Greens-plus, Greens 2.0, Off-Greens—who knows? I need to call them something for rhetorical purposes here, so I'll improvise. Combine the color green with the color of the blue sky, the blue planet, the blue ocean—all that atmospheric blue an artifact of life, back when it converted to oxygen—and what do you get? The science- and technology-loving Blue-Greens: the *Turquoise movement*, made up of Turqs (to their friends) or Turqueys (to their critics).

The Greens and Turquoises can divide up what there is to be done and still be overworked. If they maintain an ongoing, mutually respectful debate, that will help each camp critique the other's projects usefully, and they'll also know when to collaborate for focused effectiveness. (If they define themselves in partisan opposition to each other, then all is lost.)

While Greens paint the cities white and expand mass transit, pronatal Turqs work to make cities a better place for kids than the suburbs—whole neighborhoods (maybe whole high-rises) where children can walk safely to superb schools.

While Blue-Greens push for nuclear microreactors and lobby for serious fusion research, Greens hammer away on a storage technology to make wind and solar viable sources of baseload power.

While Greens ensure that the heritage of landrace variety in agricultural plants and animals is preserved, Turquoises add to that variety through genetically engineered organisms with highly specific talents and figure out how to turn algae directly into tasty food.

While Turqs restore diseased chestnut trees and extinct megafauna with engineered genomes, Greens restore carbon-absorbing tallgrass prairie and peat marshes and assemble continent-scale wildlife corridors.

To deal with climate change, while Greens pyrolize crop residue and enrich the soil with biochar, Turqueys dirty the stratosphere with sulfates. They both find ever more ingenious ways to cease using any form of combustion as a source of power.

While Greens worship Gaia, Turqs bargain with Gaia.

● The operative principle for all is what Danny Hillis calls the Golden Rule of Time: Do for the future what you're grateful the past did for you. (Or what you wish the past had done for you.) That tells you the right thing to do. As for how to do it, heed some advice from old hands:

Naturalist Peter Warshall: "Take any position and ask: *What do we want* and love? Dream the dream of the perfect (not practical) results so you can see the vision clearly and with full passion. Then ask, *What do we know?* Put together the knowledge about the situation and what facts may be missing both about the actual topic and the players and power relationships involved. Finally, *What will we accept?* You don't have to go public with your acceptance strategy, but it should be thought through."

Programmer Paul Graham: "Find (a) simple solutions (b) to overlooked problems (c) that actually need to be solved, and (d) deliver them as informally as possible, (e) starting with a very crude version 1.0, then (f) iterating rapidly." (Iterating rapidly is how squatters build cities and the Bradley sisters eliminate alien-invasive plants.)

Physicist Freeman Dyson: "A project is sustainable if it is cheap enough to be the first of a series continuing indefinitely into the future. A project is unsustainable if it is so expensive that it cannot be repeated without major political battles. A sustainable project marks the beginning of a new era. An unsustainable project marks the end of an old era."

Most innovation comes from amateurs, who are free to be radical, and from scientists in academia, who are free to follow their curiosity. But then there's a gap. It's hard to develop radical ideas into something broadly practical, because commercial money and government money are obliged to be conservative, and academic money is limited to discovery. The best money for pursuing really radical ideas into experimental use comes from individual philanthropists (foundations tend to avoid risk). According to philanthropy consultant Katherine Fulton, only one in ten wealthy people are philanthropic; 98 percent of all wealth just sits on the sidelines. Will some of that prodigious lode of cash rise to the occasion of planetary urgency? It's possible.

Governments will need encouragement to price carbon out of contention as an energy source. And governments should start running scenarios now on how they plan to deal with climate refugees.

● "It is the business of the future to be dangerous," philosopher Alfred North Whitehead said. That may account for Jim Lovelock's attitude. In this book, he is always the bearer of the bleakest news of a civilization in the process of unmaking itself, yet in person the ninety-year-old is cheerful. A reporter at the *Guardian* got him to explain why:

> Humanity is in a period exactly like 1938–9, he explains, when "we all knew something terrible was going to happen, but didn't know what to do about it." But once the second world war was under way, "everyone got excited, they loved the things they could do, it was one long holiday. . . . So when I think of

the impending crisis now, I think in those terms. A sense of purpose—that's what people want."

The shift from dread to action is under way. The outcome is wholly uncertain. Though microbes still run the world, right now they could use a little help in tuning the atmosphere. Our efforts will be tiny compared to what they do but enormous for us.

Thank you for joining those efforts, if you do. In any case, thank you for joining this discourse. I owe you a summary:

Ecological balance is too important for sentiment. It requires science.

The health of natural infrastructure is too compromised for passivity. It requires engineering.

What we call natural and what we call human are inseparable. We live one life.

Footnotes for this chapter may be found at **www.sbnotes.com**.

RECOMMENDED READING

SCIENCE

Science (weekly magazine), Bruce Alberts, editor in chief.
Nature (weekly magazine), Philip Campbell, editor in chief.
> If science is the only news, either of these prestigious weeklies will keep you current. You may want both. *Nature* is based in London, *Science* in Washington, D.C. Both publish extensive original research papers, along with articles interpreting the news, essays, book reviews, and editorials. The competition between the two magazines keeps them sharp.

New Scientist (weekly magazine), Roger Highfield, editor.
> Many magazines attempt to gauge the meaning and significance of the news coming from science, but none do it better than this weekly, also from London.

Science Daily (online), Dan and Michele Hogan, editors.
> A measure of the accelerating pace of science is the usefulness of a daily update on breaking news. This free (ad-based) Web site cov-

ers all the current science press releases six times a day. It's easy to tune your news feeds for just the subjects you're interested in.

SciDev.Net (online), David Dickson, director.
Applications of new science and technology for the developing world are the focus of this remarkable site. No ads; it is supported by aid organizations. By arrangement with *Science* and *Nature*, links to full-length articles and papers in those publications can be followed without paying their subscription fee.

Technology Review (bimonthly magazine), Jason Pontin, editor in chief.
With its lively new Web site, this is now the best of many publications that track new technologies. Its editor and authors are comfortable voicing strong opinions.

National Geographic (monthly magazine), Chris Johns, editor in chief.
No other magazine has maintained such high quality for so long to such good effect. The world's finest photographs and graphics are now being matched with excellent writing to present current science in a planetary context.

CLIMATE

The Revenge of Gaia: Why the Earth Is Fighting Back—and How We Can Still Save Humanity (2006), James Lovelock.
The Vanishing Face of Gaia: A Final Warning (2009), James Lovelock.
Of the many scientists studying climate change, Lovelock has the most comprehensive contemporary perspective, thanks to his decades of work on Gaia theory. These two books detail his analysis of how extreme the situation is becoming.

Six Degrees: Our Future on a Hotter Planet (2007), Mark Lynas.
Lynas succeeds where most others fail in making inescapably clear

how increasingly inhospitable the world will be with each increase of global temperature from 1° to 6°C. The book is a cure for an incrementalist approach to climate change. You don't think "We can handle a 2-degree rise" after you learn what that will mean.

Plows, Plagues, and Petroleum: How Humans Took Control of Climate (2005), William Ruddiman.
   If Ruddiman is right, climate is extraordinarily sensitive to human activity. That might be good news.

Climate Debate Daily (online), Douglas Campbell and Denis Dutton, editors.
   There are scores of Web sites collecting science news, political news, blog commentaries, etc., related to climate. Some are alarmist, some are skeptical, some are neutral. This one is aggressively and fascinatingly neutral, presenting a rich brew of strong comments on every climate topic in a debatelike format. The hosts are philosophy professors.

CITIES

Shack/Slum Dwellers International (online)
   With SDI, the world's urban poor have their own comprehensive Web site, linking the slum-improvement activities in thirty countries of the global south. Click around in it for a tour of amazing activities.

Shantaram: A Novel (2005), Gregory David Roberts.
   This is the *Les Misérables* of the twenty-first century. Set in Mumbai's slums and underworld, it is one of the great romances.

The Places We Live (2008), Jonas Bendiksen.
   First-rate photojournalism. Like the renowned Farm Security Administration photographers of the American Depression

years, Bendiksen went to the slums of Caracas, Nairobi, Mumbai, and Jakarta and brought back beautiful photographs from inside the shacks, along with first-person accounts by the shack dwellers.

Shadow Cities: A Billion Squatters, a New Urban World (2004), Robert Neuwirth.
Neuwirth demonstrates adventurous journalism at its best. When you're curious or worried about something, go and live there. You'll learn what to be really worried about and what to be inspired by.

Squatter City (online); Stealth of Nations (online), Robert Neuwirth, blogger.
Neuwirth is working on a book about the informal economy in slums and elsewhere. His two blogs keep you current on what he's finding.

Unleashing the Potential of Urban Growth: State of World Population 2007, United Nations Population Fund.
An upbeat, realistic survey of the current state of urbanization. It is a textbook for sensible urban policy.

Infrastructure: A Field Guide to the Industrial Landscape (2005), Brian Hayes.
We know more about the things on trees (leaves, twigs, bark) than about the things on telephone poles (primary and secondary electricity distribution lines, insulators, switches, fuses, transformers, street lights, cable TV feeders, phone cables, and the grounding lead). This glorious book cures ignorance on every infrastructural subject.

POPULATION

The Empty Cradle: How Falling Birthrates Threaten World Prosperity and What to Do About It (2004), Phillip Longman.

Fewer: How the New Demography of Depopulation Will Shape Our Future (2004), Ben Wattenberg.

Longman is a liberal, Wattenberg a conservative. Wattenberg is the better writer; Longman will get a more sympathetic hearing with environmentalists. Both are rightly worried.

NUCLEAR

Power to Save the World: The Truth About Nuclear Energy (2007), Gwyneth Cravens.

Besides presenting a persuasive case for nuclear power, the book is an exemplary account of a Green coming to see the world the way an engineer does. It demonstrates why more should.

NEI Nuclear Notes (online), Mark Flanagan and David Bradish, lead bloggers.

Nuclear advocacy at its best. Open-minded, even-handed, and alert, the Nuclear Energy Institute bloggers give context to nuclear news in a way that mainstream media don't.

GENETIC ENGINEERING

Tomorrow's Table: Organic Farming, Genetics, and the Future of Food (2008), Pamela Ronald and Raoul Adamchak.

Organic farming marries genetic engineering and lives happily ever after. The book has a real-life texture missing in most works about GE or organic.

The Doubly Green Revolution: Food for All in the Twenty-first Century (1999), Gordon Conway.

Experience tells. Conway has seen it all and knows exactly how GE fits into simultaneously feeding the world and protecting the environment.

Starved for Science: How Biotechnology Is Being Kept Out of Africa (2008), Robert Paarlberg.

　　Anatomy of an ongoing Green-sponsored atrocity in Africa.

Mendel in the Kitchen: A Scientist's View of Genetically Modified Foods (2006), Nina Fedoroff.

　　Geneticist Federoff gives a much fuller background for how GE works with food crops than I could.

CropBiotech Update (online).

　　The successes of GE throughout the world, along with entanglements it meets, are chronicled on a daily basis here.

New Science of Metagenomics: Revealing the Secrets of Our Microbial Planet (2007), Board on Life Sciences.

　　The oldest and by far the most profuse form of life is finally being studied properly, and we learn we've been living all this time on "the planet of the bacteria."

ENVIRONMENTALISM

Conservation (quarterly), Kathryn Kohm, editor.

　　Good editors challenge their readers. Kohm publishes heresies and innovations as well as the workaday discoveries in conservation biology. The result is an environmental publication filled with real news.

High Country News (biweekly), Jonathan Thompson, editor.

　　This tabloid specializes in Green reporting with a genuinely neutral point of view so that ranchers, loggers, hunters, businesspeople, and bureaucrats feel included rather than demonized. The region served is the American West.

Sierra (bimonthly), Bob Sipchen, editor in chief.

A full-feature environmentalist magazine, this publication from the Sierra Club focuses on practical matters rather than on the contentless inspirational woo-woo typical of Green periodicals such as *Orion*.

Earth Island Journal (quarterly), Jason Mark, editor.
For a thoroughly partisan Green publication, this one has impressive journalistic reach.

OnEarth (quarterly), Douglas Barasch, editor in chief.
Like its parent, the Natural Resources Defense Council, *OnEarth* is sober, careful, wide ranging, and not allergic to business or government in its Green advocacy.

GreenFacts (online), Jacques Wirtgen, general manager.
Based in Brussels, GreenFacts "provide summaries of scientific consensus reports on environment and health issues" in English, French, Spanish, and German. Its funding comes from nonprofits and governments as well as from private companies. The summaries are thorough, admirably designed for the Web, and linked to original sources. If you're researching any Green subject, this is the first place to check.

Constant Battles: The Myth of the Peaceful, Noble Savage (2003), Steven LeBlanc with Katherine Register.
There is a lot more to LeBlanc's book than what I've summarized. Nature is best understood if you include the harsh parts and the same is true of humanity.

The Idea of Decline in Western History (1997), Arthur Herman.
Romanticism has left a trail of bodies, many of them suicides, ever since Rousseau. It is a cult of heroic despair that ill serves the environmental movement.

Blessed Unrest: How the Largest Movement in the World Came into Being and Why No One Saw It Coming (2007), Paul Hawken.

This is the closest we have to a *Whole Earth Catalog* of environmental and social justice organizations. I would love to see its online expression, WiserEarth.org, become truly comprehensive.

Counterculture Green: The Whole Earth Catalog and American Environmentalism (2007), Andrew Kirk.
Some of the origins of the book you're holding can be traced in Kirk's study of the Green influence of the original *Whole Earth Catalog*.

Earthrise: How Man First Saw the Earth (2008), Robert Poole.
Don't take my word that the first Earth photographs were a boon for environmentalists. Poole chronicles the whole original event and the worldwide inspiration that resulted.

ECOLOGY

The Future of Life (2002), Edward O. Wilson.
Naturalist (1994), Edward O. Wilson.
Wilson's memoir, *Naturalist*, details one of the most productive lives in science and *The Future of Life* lays out the road map of needs and techniques for conservation biology worldwide.

The Wild Trees: A Story of Passion and Daring (2007), Richard Preston.
I wanted to include in *Discipline* a paean to the intrepid lives of field biologists, but I couldn't fit it in. This book will do as a prime sample. In the tops of the tallest trees in the world a whole ecosystem was discovered by acrobatic biologists. *New Yorker* writer Preston learned the death-defying skills to join them there.

Lament for an Ocean: The Collapse of the Atlantic Cod Fishery (1998), Michael Harris.
What happens when you try to protect the fishermen rather than the fish.

Degrees of Disaster: Prince William Sound: How Nature Reels and Rebounds (1996), Jeff Wheelwright.

>   Deliciously inconvenient ground truth at the site of one of the great ecological-political arguments.

Sustaining Life: How Human Health Depends on Biodiversity (2008), Eric Chivian and Aaron Bernstein, editors.

>   An exploration, with dazzling graphics, of one of the most essential ecosystem services.

Ocean: An Illustrated Atlas (2008), Sylvia Earle and Linda Glover.

>   The best compendium of recent discoveries about the oceans and ocean life, presented with *National Geographic* panache. For humans this may be a "city planet," but for life and climate it's an ocean planet.

INDIANS

1491: New Revelations of the Americas Before Columbus (2005), Charles C. Mann.

>   What really happened on our continent is totally different from what we learned in school.

Tending the Wild: Native American Knowledge and the Management of California's Natural Resources (2005), M. Kat Anderson.

>   I wish every region in the world could have so complete an account of how the first human inhabitants engineered the ecosystem.

Where the Lightning Strikes: The Lives of American Indian Sacred Places (2007), Peter Nabokov.

>   The land is alive with ancient attention and reverence, and that continuity is worth maintaining.

RESTORATION

Bringing Back the Bush: The Bradley Method of Bush Regeneration (2002), Joan Bradley. (Most easily purchased online from CSIRO Publishing.)
>Patience, vigilance, subtlety, craft, and success characterize the Bradley sisters' approach to defeating alien-invasive plants. The book is specific to Australia, but its techniques apply everywhere.

Nature by Design: People, Natural Process, and Ecological Restoration (2003), Eric Higgs.
>Higgs is the first to offer a compelling general theory of restoration, leading to intelligent policy and practices.

Green Phoenix: Restoring the Tropical Forests of Guanacaste, Costa Rica (2003), William Allen.
>Preservation in the real world is always a richly tangled tale. Here is one of the great stories, with charismatic, eloquent Daniel Janzen in the middle of it.

Where the Land Is Greener: Case Studies and Analysis of Soil and Water Conservation Initiatives Worldwide (2007), Hanspeter Liniger, editor.
>There are so many ingenious ways to bring life and productivity back to degraded land. Collecting them all in one book is a tremendous public service.

Where the Wild Things Were: Life, Death, and Ecological Wreckage in a Land of Vanishing Predators (2008), William Stolzenburg.
>A well-written and persuasive presentation of essential-predator theory.

Life Out of Bounds: Bioinvasion in a Borderless World (1998), Chris Bright.
>This is the best survey I've seen on the impacts of alien-invasive species and what to do about them.

The World Without Us (2007), Alan Weisman.
   Exceptionally thorough field research distinguishes this account of what life gets up to as soon as humans step away. It is a fascinating read.

GEOENGINEERING

Dune (science fiction, 1965), Frank Herbert.
   Because there were no nonfiction books on geoengineering available when I finished *Discipline* (though Jeff Goodell was working on one), I was forced to look for the cream of science fiction terra-forming stories. *Dune* is as creamy as it gets. It features a planetary ecologist inspiring a native people to subversively convert their desert planet to an oasis.

Red Mars (1992), *Green Mars* (1993), *Blue Mars* (1996), science fiction trilogy by Kim Stanley Robinson.
   This is a fine "hard science" saga of the human drama and climatological sophistication it might take to make Mars a good place to live. I could imagine a similar "Blue Earth," "Brown Earth" sequence of stories examining what would happen here if the climate tips all the way out of Gaian control and Earth becomes the next Mars.

ACKNOWLEDGMENTS

Magazine editors cause books. In my case two new-on-the-job editors are to blame. Jason Pontin, who used to sweat with me in a mountain fitness program, was hired as editor of MIT's *Technology Review* in 2004. Art Kleiner, a longtime *Whole Earth* editor, took over the editorship of Booz Allen Hamilton's business quarterly, *Strategy + Business*, in 2005. New editors are supposed to bring new writers, so both invited me to submit something. For Pontin I wrote a brief polemic titled "Environmental Heresies." Kleiner got a long-form article titled "City Planet." The two pieces stirred up interest, which led to some secondary press, and that led to a book proposal and contract, and here we are.

Literary agent John Brockman is another book causer. Every book I've worked on since 1972 has conduited through him. During that time he mustered the world's finest collection of scientist writers, engineered horizortal idea flow among them, nagged them toward publication, and enriched them. It's a family business. John's wife Katinka Madsen and son Max Brockman were also in the thick of this project.

Drafts of my chapters were vetted by Paul Slovak, John Brockman, Nils Gilman, Robert Fuller, Brian Eno, Kevin Kelly, Ryan Phelan, Alexander Rose, George Dyson, James Lovelock, Richard Rhodes, Gwyneth

Cravens, Rip Anderson, Peter Schwartz, Daniel Janzen, Pamela Ronald, Raoul Adamchak, José Baer, Peter Raven, Rob Carlson, and Rusty Schweickart. I hired James Donnelly for a first round of copyediting, and freelancer Gary Stimeling did the second round. Editor Paul Slovak ran the show at the Viking Penguin end. Other credits there include: jacket design, Gregg Kulik; interior design, Ginger Legato; index, **TK**. For the UK-based Grove-Atlantic edition, Toby Mundy was editor, and others involved were: **TK**.

Throughout *Discipline* I make a number of predictions for which I should be held accountable. They include forecasts about urbanization after the 2009 financial crisis, peak world population, Chernobyl National Park, GE poplars in China, the future opinions of Amory Lovins and Bjørn Lomborg, the engineered revival of the American chestnut, and GE biocontrol organisms for restoration. Formal, falsifiable versions of the predictions, complete with my detailed argument in each case, have been placed on a Web site called Long Bets: www.longbets.org. There you may vote on my forecasts, comment about them, and even bet money against me about them. As history proceeds, you can watch me be wrong, or maybe even right. Better still, post your own predictions for the judgment of history.

—Stewart Brand
April 2009